Statistical Techniques

for

Data Analysis

Second Edition

Statistical Techniques

for

Data Analysis

Second Edition

John K. Taylor Ph.D.

**Formerly of the National Institute of
Standards and Technology**

and

Cheryl Cihon Ph.D.

Bayer HealthCare, Pharmaceuticals

CHAPMAN & HALL/CRC

A CRC Press Company

Boca Raton   London   New York   Washington, D.C.

Library of Congress Cataloging-in-Publication Data

Cihon, Cheryl.
    Statistical techniques for data analysis / Cheryl Cihon, John K. Taylor.—2nd. ed.
        p. cm.
    Includes bibliographical references and index.
    ISBN 1-58488-385-5 (alk. paper)
    1. Mathematical statistics. I. Taylor, John K. (John Keenan), 1912-II. Title.

QA276.C4835 2004
519.5—dc22
                                                                      2003062744

Visit the CRC Press Web site at www.crcpress.com

© 2004 by Chapman & Hall/CRC

No claim to original U.S. Government works
International Standard Book Number 1-58488-385-5
Library of Congress Card Number 2003062744
Printed in the United States of America 1 2 3 4 5 6 7 8 9 0
Printed on acid-free paper

Preface

Data are the products of measurement. Quality measurements are only achievable if measurement processes are planned and operated in a state of statistical control. Statistics has been defined as the branch of mathematics that deals with all aspects of the science of decision making in the face of uncertainty. Unfortunately, there is great variability in the level of understanding of basic statistics by both producers and users of data.

The computer has come to the assistance of the modern experimenter and data analyst by providing techniques for the sophisticated treatment of data that were unavailable to professional statisticians two decades ago. The days of laborious calculations with the ever-present threat of numerical errors when applying statistics of measurements are over. Unfortunately, this advance often results in the application of statistics with little comprehension of meaning and justification. Clearly, there is a need for greater statistical literacy in modern applied science and technology.

There is no dearth of statistics books these days. There are many journals devoted to the publication of research papers in this field. One may ask the purpose of this particular book. The need for the present book has been emphasized to the authors during their teaching experience. While an understanding of basic statistics is essential for planning measurement programs and for analyzing and interpreting data, it has been observed that many students have less than good comprehension of statistics, and do not feel comfortable when making simple statistically based decisions. One reason for this deficiency is that most of the numerous works devoted to statistics are written for statistically informed readers.

To overcome this problem, this book is not a statistics textbook in any sense of the word. It contains no theory and no derivation of the procedures presented and presumes little or no previous knowledge of statistics on the part of the reader. Because of the many books devoted to such matters, a theoretical presentation is deemed to be unnecessary, However, the author urges the reader who wants more than a working knowledge of statistical techniques to consult such books. It is modestly hoped that the present book will not only encourage many readers to study statistics further, but will provide a practical background which will give increased meaning to the pursuit of statistical knowledge.

This book is written for those who make measurements and interpret experimental data. The book begins with a general discussion of the kinds of data and how to obtain meaningful measurements. General statistical principles are then described, followed by a chapter on basic statistical calculations. A number of the most frequently used statistical techniques are described. The techniques are arranged for presentation according to decision situations frequently encountered in measurement or data analysis. Each area of application and corresponding technique is explained in general terms yet in a correct scientific context. A chapter follows that is devoted to management of data sets. Ways to present data by means of tables, charts, graphs, and mathematical expressions are next considered. Types of data that are not continuous and appropriate analysis techniques are then discussed. The book concludes with a chapter containing a number of special techniques that are used less frequently than the ones described earlier, but which have importance in certain situations.

Numerous examples are interspersed in the text to make the various procedures clear. The use of computer software with step-by-step procedures and output are presented. Relevant exercises are appended to each chapter to assist in the learning process.

The material is presented informally and in logical progression to enhance readability. While intended for self-study, the book could provide the basis for a short course on introduction to statistical analysis or be used as a supplement to both undergraduate and graduate studies for majors in the physical sciences and engineering.

The work is not designed to be comprehensive but rather selective in the subject matter that is covered. The material should pertain to most everyday decisions relating to the production and use of data.

Acknowledgments

The second author would like to express her gratitude to all the teachers of statistics who, over the years, encouraged her development in the area and gave her the tools to undertake such a project.

Dedication

This book is dedicated to the husband, son and family of Cheryl A. Cihon, and to the memory of John K. Taylor.

The late John K. Taylor was an analytical chemist of many years of varied experience. All of his professional life was spent at the National Bureau of Standards, now the National Institute of Standards and Technology, from which he retired after 57 years of service.

Dr. Taylor received his BS degree from George Washington University and MS and PhD degrees from the University of Maryland. At the National Bureau of Standards, he served first as a research chemist, and then managed research and development programs in general analytical chemistry, electrochemical analysis, microchemical analysis, and air, water, and particulate analysis. He coordinated the NBS Center for Analytical Chemistry's Program in quality assurance, and conducted research activities to develop advanced concepts to improve and assure measurement reliability. He provided advisory services to other government agencies as part of his official duties as well as consulting services to government and industry in analytical and measurement programs.

Dr. Taylor authored four books, and wrote over 220 research papers in analytical chemistry. Dr. Taylor received several awards for his accomplishments in analytical chemistry, including the Department of Commence Silver and Gold Medal Awards. He served as past chairman of the Washington Academy of Sciences, the ACS Analytical Chemistry Division, and the ASTM Committee D 22 on Sampling and Analysis of Atmospheres.

Cheryl A. Cihon is currently a biostatistician in the pharmaceutical industry where she works on drug development projects relating to the statistical aspects of clinical trial design and analysis.

Dr. Cihon received her BS degree in Mathematics from McMaster University, Ontario, Canada as well as her MS degree in Statistics. Her PhD degree was granted from the University of Western Ontario, Canada in the field of Biostatistics. At the Canadian Center for Inland Waters, she was involved in the analysis of environmental data, specifically related to toxin levels in major lakes and rivers throughout North America. Dr. Cihon also worked as a statistician at the University of Guelph, Canada, where she was involved with analyses pertaining to population medicine. Dr. Cihon has taught many courses in advanced statistics throughout her career and served as a statistical consultant on numerous projects.

Dr. Cihon has authored one other book, and has written many papers for statistical and pharmaceutical journals. Dr. Cihon is the recipient of several awards for her accomplishments in statistics, including the National Sciences and Engineering Research Council award.

Table of Contents

Preface v

CHAPTER 1. What Are Data? 1
  Definition of Data 1
  Kinds of Data 2
    Natural Data 2
    Experimental Data... 3
      Counting Data and Enumeration 3
      Discrete Data 4
      Continuous Data 4
  Variability 4
  Populations and Samples... 5
  Importance of Reliability 5
  Metrology.. 6
  Computer Assisted Statistical Analyses 7
  Exercises 8
  References.. 8

CHAPTER 2. Obtaining Meaningful Data 10
  Data Production Must Be Planned 10
  The Experimental Method... 11
    What Data Are Needed.. 12
    Amount of Data... 13
    Quality Considerations 13
  Data Quality Indicators 13
  Data Quality Objectives 15
  Systematic Measurement 15
  Quality Assurance 15
  Importance of Peer Review... 16
  Exercises 17
  References.. 17

CHAPTER 3. General Principles.. 19
    Introduction.. 19
    Kinds of Statistics 20
    Decisions.. 21
    Error and Uncertainty.. 22
    Kinds of Data 22
    Accuracy, Precision, and Bias... 22
    Statistical Control.. 25
        Data Descriptors.. 25
    Distributions... 27
    Tests for Normality 30
    Basic Requirements for Statistical Analysis Validity...................................... 36
    MINITAB 39
        Introduction to MINITAB.. 39
        MINITAB Example 42
    Exercises 44
    References... 45

CHAPTER 4. Statistical Calculations.. 47
    Introduction.. 47
    The Mean, Variance, and Standard Deviation... 48
    Degrees of Freedom 52
    Using Duplicate Measurements to Estimate a Standard Deviation 52
    Using the Range to Estimate the Standard Deviation 54
    Pooled Statistical Estimates 55
    Simple Analysis of Variance... 56
    Log Normal Statistics.. 64
    Minimum Reporting Statistics 65
    Computations 66
    One Last Thing to Remember 68
    Exercises 68
    References... 71

CHAPTER 5. Data Analysis Techniques.. 72
    Introduction.. 72
    One Sample Topics 73
        Means 73
            Confidence Intervals for One Sample.. 73
            Does a Mean Differ Significantly from a Measured or Specified Value.. 77
        MINITAB Example... 78
        Standard Deviations 80

Confidence Intervals for One Sample.. 80
Does a Standard Deviation Differ Significantly from a
Measured or Specified Value.. 81
MINITAB Example... 82
Statistical Tolerance Intervals 82
Combining Confidence Intervals and Tolerance Intervals 85
Two Sample Topics 87
Means 87
Do Two Means Differ Significantly 87
MINITAB Example... 90
Standard Deviations 91
Do Two Standard Deviations Differ Significantly 91
MINITAB Example... 93
Propagation of Error in a Derived or Calculated Value 94
Exercises 96
References.. 99

CHAPTER 6. Managing Sets of Data... 100
Introduction.. 100
Outliers... 100
The Rule of the Huge Error 101
The Dixon Test.. 102
The Grubbs Test.. 104
Youden Test for Outlying Laboratories 105
Cochran Test for Extreme Values of Variance............................. 107
MINITAB Example... 108
Combining Data Sets... 109
Statistics of Interlaboratory Collaborative Testing............................. 112
Validation of a Method of Test 112
Proficiency Testing... 113
Testing to Determine Consensus Values of Materials.............. 114
Random Numbers 114
MINITAB Example... 115
Exercises 118
References.. 120

CHAPTER 7. Presenting Data... 122
Tables... 122
Charts 123
Pie Charts 123
Bar Charts.. 123
Graphs.. 126
Linear Graphs.. 126
Nonlinear Graphs 127
Nomographs 128
MINITAB Example... 128

Mathematical Expressions 131
  Theoretical Relationships 131
  Empirical Relationships 132
    Linear Empirical Relationships 132
    Nonlinear Empirical Relationships... 133
    Other Empirical Relationships... 133
  Fitting Data.. 133
    Method of Selected Points.. 133
    Method of Averages 134
    Method of Least Squares 137
  MINITAB Example... 140
  Summary 143
Exercises 144
References.. 145

CHAPTER 8. Proportions, Survival Data and Time Series Data 147
  Introduction... 147
  Proportions.. 148
    Introduction 148
    One Sample Topics 148
      Two-Sided Confidence Intervals for One Sample 149
      MINITAB Example.. 150
      One-Sided Confidence Intervals for One Sample................. 150
      MINITAB Example.. 151
      Sample Sizes for Proportions-One Sample........................... 152
      MINITAB Example.. 153
    Two Sample Topics... 153
      Two-Sided Confidence Intervals for Two Samples............... 154
      MINITAB Example.. 154
      Chi-Square Tests of Association 155
      MINITAB Example.. 156
      One-Sided Confidence Intervals for Two Samples 157
      Sample Sizes for Proportions-Two Samples.......................... 157
      MINITAB Example.. 158
  Survival Data... 159
    Introduction 159
    Censoring 159
    One Sample Topics 160
      Product Limit/Kaplan Meier Survival Estimate 161
      MINITAB Example.. 162
    Two Sample Topics... 165
      Proportional Hazards 165
      Log Rank Test 165
      MINITAB Example.. 169
      Distribution Based Survival Analyses 170
      MINITAB Example.. 170

Summary 174
Time Series Data 174
  Introduction 174
  Data Presentation 175
    Time Series Plots 176
    MINITAB Example 176
    Smoothing 177
    MINITAB Example 178
    Moving Averages 180
    MINITAB Example 181
  Summary 181
  Exercises 182
  References 184

CHAPTER 9. Selected Topics 185
  Basic Probability Concepts 185
  Measures of Location 187
    Mean, Median, and Midrange 187
    Trimmed Means 188
      Average Deviation 188
  Tests for Nonrandomness 189
    Runs 190
      Runs in a Data Set 190
      Runs in Residuals from a Fitted Line 191
    Trends/Slopes 191
    Mean Square of Successive Differences 192
  Comparing Several Averages 194
  Type I Errors, Type II Errors and Statistical Power 195
    The Sign of the Difference is Not Important 197
    The Sign of the Difference is Important 198
    Use of Relative Values 199
    The Ratio of Standard Deviation to Difference 199
  Critical Values and P Values 200
    MINITAB Example 201
  Correlation Coefficient 206
    MINITAB Example 209
  The Best Two Out of Three 209
  Comparing a Frequency Distribution with a Normal Distribution 210
  Confidence for a Fitted Line 211
    MINITAB Example 215
  Joint Confidence Region for the Constants of a Fitted Line 215
  Shortcut Procedures 216
  Nonparametric Tests 217
    Wilcoxon Signed-Rank Test 217
    MINITAB Example 220

Extreme Value Data 220
Statistics of Control Charts 221
   Property Control Charts.. 221
   Precision Control Charts 223
   Systematic Trends in Control Charts.. 224
Simulation and Macros 224
      MINITAB Example... 225
   Exercises 226
   References.. 229

CHAPTER 10. Conclusion 231
   Summary... 231

Appendix A. Statistical Tables 233

Appendix B. Glossary... 244

Appendix C. Answers to Numerical Exercises 254

Index 269

List of Figures

Figure 1.1   Role of statistics in metrology 7
Figure 3.1   Measurement decision 21
Figure 3.2   Types of data 23
Figure 3.3   Precision and bias 24
Figure 3.4   Normal distribution 28
Figure 3.5   Several kinds of distributions 29
Figure 3.6   Variations of the normal distribution 30
Figure 3.7   Histograms of experimental data 31
Figure 3.8   Normal probability plot 34
Figure 3.9   Log normal probability plot 35
Figure 3.10   Log × normal probability plot 36
Figure 3.11   Probability plots 37
Figure 3.12   Skewness 38
Figure 3.13   Kurtosis 39
Figure 3.14   Experimental uniform distribution 40
Figure 3.15   Mean of ten casts of dice 40
Figure 3.16   Gross deviations from randomness 41
Figure 3.17   Normal probability plot-membrane method 44
Figure 4.1   Population values and sample estimates 49
Figure 4.2   Distribution of means 50
Figure 5.1   90% confidence intervals 76
Figure 5.2   Graphical summary including confidence interval for standard
              deviation 83
Figure 5.3   Combination of confidence and tolerance intervals 87
Figure 5.4   Tests for equal variances 94
Figure 6.1   Boxplot of titration data 109
Figure 6.2   Combining data sets 111
Figure 7.1   Typical pie chart 124
Figure 7.2   Typical bar chart 125
Figure 7.3   Pie chart of manufacturing defects 129
Figure 7.4   Linear graph of cities data 130
Figure 7.5   Linear graph of cities data-revised 131
Figure 7.6   Normal probability plot of residuals 141
Figure 8.1   Kaplan Meier survival plot 164

Figure 8.2   Survival distribution identification 172
Figure 8.3   Comparing log normal models for reliable dataset 173
Figure 8.4   Time series plot 178
Figure 8.5   Smoothed time series plot 180
Figure 8.6   Moving averages of crankshaft dataset 182
Figure 9.1   Critical regions for 2-sided hypothesis tests 202
Figure 9.2   Critical regions for 1-sided upper hypothesis tests 202
Figure 9.3   Critical regions for 1-sided lower hypothesis tests 203
Figure 9.4   P value region 204
Figure 9.5   OC curve for the two-sided t test ($\alpha = .05$) 207
Figure 9.6   Superposition of normal curve on frequency plot 212
Figure 9.7   Calibration data with confidence bands 215
Figure 9.8   Joint confidence region ellipse for slope and intercept of a
               linear relationship 218
Figure 9.9   Maximum tensile strength of aluminum alloy 222

List of Tables

Table 2.1. Items for Consideration in Defining a Problem for
            Investigation 11
Table 3.1. Limits for the Skewness Factor, g_1, in the Case of a
            Normal Distribution 38
Table 3.2. Limits for the Kurtosis Factor, g_2, in the Case of a
            Normal Distribution 39
Table 3.3. Radiation Dataset from MINITAB 42
Table 4.1. Format for Tabulation of Data Used in Estimation of Variance
            at Three Levels, Using a Nested Design Involving Duplicates 62
Table 4.2. Material Bag Dataset from MINITAB 63
Table 5.1. Furnace Temperature Dataset from MINITAB 78
Table 5.2. Comparison of Confidence and Tolerance Interval Factors 85
Table 5.3. Acid Dataset from MINITAB 90
Table 5.4. Propagation of Error Formulas for Some Simple Functions 95
Table 6.1. Random Number Distributions 116
Table 7.1. Some Linearizing Transformations 127
Table 7.2. Cities Dataset from MINITAB 130
Table 7.3. Normal Equations for Least Squares Curve Fitting for the
            General Power Series $Y = a + bX + cX^2 + dX^3 +$ 136
Table 7.4. Normal Equations for Least Squares Curve Fitting for the Linear
            Relationship $Y = a + bX$ 136
Table 7.5. Basic Worksheet for All Types of Linear Relationships 138
Table 7.6. Furnace Dataset from MINITAB 140
Table 8.1. Reliable Dataset from MINITAB 162
Table 8.2. Kaplan Meier Calculation Steps 163
Table 8.3. Log Rank Test Calculation Steps 167
Table 8.4. Crankshaft Dataset from MINITAB 176
Table 8.5. Crankshaft Dataset Revised 177
Table 8.6. Crankshaft Means by Time 177
Table 9.1. Ratio of Average Deviation to Sigma for Small Samples 189
Table 9.2. Critical Values for the Ratio MSSD/Variance 193
Table 9.3. Percentiles of the Studentized Range, $q_{.95}$ 194
Table 9.4. Sample Sizes Required to Detect Prescribed Differences
            between Averages when the Sign Is Not Important 198

Table 9.5. Sample Sizes Required to Detect Prescribed Differences
             between Averages when the Sign Is Important................................ 199
Table 9.6. 95% Confidence Belt for Correlation Coefficient............................. 208
Table 9.7. Format for Use in Construction of a Normal Distribution 210
Table 9.8. Normalization Factors for Drawing a Normal Distribution 211
Table 9.9. Values for $F_{1-\alpha}$ ($\alpha = .95$) for ($2, n - 2$) 213
Table 9.10. Wilcoxon Signed-Rank Test Calculations 219
Table 9.11. Control Chart Limits 223

CHAPTER 1

What are Data?

Data may be considered to be one of the vital fluids of modern civilization. Data are used to make decisions, to support decisions already made, to provide reasons why certain events happen, and to make predictions on events to come. This opening chapter describes the kinds of data used most frequently in the sciences and engineering and describes some of their important characteristics.

DEFINITION OF DATA

The word data is defined as things known, or assumed facts and figures, from which conclusions can be inferred. Broadly, data is raw information and this can be qualitative as well as quantitative. The source can be anything from hearsay to the result of elegant and painstaking research and investigation. The terms of reporting can be descriptive, numerical, or various combinations of both. The transition from data to knowledge may be considered to consist of the hierarchal sequence

$$\text{Data} \xrightarrow{\text{analysis}} \text{Information} \xrightarrow{\text{model}} \text{Knowledge}$$

Ordinarily, some kind of analysis is required to convert data into information. The techniques described later in this book often will be found useful for this purpose. A model is typically required to interpret numerical information to provide knowledge about a specific subject of interest. Also, data may be acquired, analyzed, and used to test a model of a particular problem.

Data often are obtained to provide a basis for decision, or to support a decision that may have been made already. An objective decision requires unbiased data but this should never be assumed. A process used for the latter purpose may be more biased than one for the former purpose, to the extent that the collection, accumulation, or production process may be biased, which is to say it may ignore other possible bits of information. Bias may be accidental or intentional. Preassumptions and even prior misleading data can be responsible for intentional bias, which may be justified. Unfortunately, many compilations of data provide little if any information about intentional biases or modifying circumstances that could affect decisions based upon them, and certainly nothing about unidentified bias.

Data producers have the obligation to present all pertinent information that would impact on the use of it, to the extent possible. Often, they are in the best position to provide such background information, and they may be the only source of information on these matters. When they cannot do so, it may be a condemnation of their competence as metrologists. Of course, every possible use of data cannot be envisioned when it is produced, but the details of its production, its limitations, and quantitative estimates of its reliability always can be presented. Without such, data can hardly be classified as useful information.

Users of data cannot be held blameless for any misuse of it, whether or not they may have been misled by its producer. No data should be used for any purpose unless their reliability is verified. No matter how attractive it may be, unevaluated data are virtually worthless and the temptation to use them should be resisted. Data users must be able to evaluate all data that they utilize or depend on reliable sources to provide such information to them.

It is the purpose of this book to provide insight into data evaluation processes and to provide guidance and even direction in some situations. However, the book is not intended and cannot hope to be used as a "cook book" for the mechanical evaluation of numerical information.

KINDS OF DATA

Some data may be classified as "soft" which usually is qualitative and often makes use of words in the form of labels, descriptors, or category assignments as the primary mode of conveying information. Opinion polls provide soft data, although the results may be described numerically. Numerical data may be classified as "hard" data, but one should be aware, as already mentioned, that such can have a soft underbelly. While recognizing the importance of soft data in many situations, the chapters that follow will be concerned with the evaluation of numerical data. That is to say, they will be concerned with quantitative, instead of qualitative data.

Natural Data

For the purposes of the present discussion, natural data is defined as that describing natural phenomena, as contrasted with that arising from experimentation. Observations of natural phenomena have provided the background for scientific theory and principles and the desire to obtain better and more accurate observations has been the stimulus for advances in scientific instrumentation and improved methodology. Physical science is indebted to natural science which stimulated the development of the science of statistics to better understand the variability of nature. Experimental studies of natural processes provided the impetus for the development of the science of experimental design and planning. The boundary between physical and natural science hardly exists anymore, and the latter now makes extensive use of physical measuring techniques, many of which are amenable to the data evaluation procedures described later.

Studies to evaluate environmental problems may be considered to be studies of natural phenomena in that the observer plays essentially a passive role. However, the observer can have control of the sampling aspects and should exercise it, judiciously, to obtain meaningful data.

Experimental Data

Experimental data result from a measurement process in which some property is measured for characterization purposes. The data obtained consist of numbers that often provide a basis for decision. This can range anywhere from discarding the data, modifying it by exclusion of some point or points, or using it alone or in connection with other data in a decision process. Several kinds of data may be obtained as will be described below.

Counting Data and Enumeration

Some data consist of the results of counting. Provided no blunders are involved, the number obtained is exact. Thus several observers would be expected to obtain the same result. Exceptions would occur when some judgment is involved as to what to count and what constitutes a valid event or an object that should be counted. The optical identification and counting of asbestos fibers is an example of the case in point. Training of observers can minimize variability in such cases and is often required if consistency of data is to be achieved. Training is best done on a direct basis, since written instructions can be subject to variable interpretation. Training often reflects the biases of the trainer. Accordingly, serial training (training some one who trains another who, in turn, trains others) should be avoided. Perceptions can change with time, in which case training may need to be a continuing process. Any process involving counting should not be called measurement but rather enumeration.

Counting of radioactive disintegrations is a special and widely practiced area of counting. The events counted (e.g., disintegrations) follow statistical principles that are well understood and used by the practitioners, so will not be discussed here. Experimental factors such as geometric relations of samples to counters and the efficiency of detectors can influence the results, as well. These, together with sampling, introduce variability and sources of bias into the data in much the same way as happens for other types of measurement and thus can be evaluated using the principles and practices discussed here.

Discrete Data

Discrete data describes numbers that have a finite possible range with only certain individual values encountered within this range. Thus, the faces on a die can be numbered, one to six, and no other value can be recorded when a certain face appears.

Numerical quantities can result from mathematical operations or from measurements. The rules of significant figures apply to the former and statistical significance applies to the latter. Trigonometric functions, logarithms, and the value of π, for example, have discrete values but may be rounded off to any number of figures for computational or tabulation purposes. The uncertainty of such numbers is due to rounding alone, and is quite a different matter from measurement uncertainty. Discrete numbers should be used in computation, rounded consistent with the experimental data to which they relate, so that the rounding does not introduce significant error in a calculated result.

Continuous Data

Measurement processes usually provide continuous data. The final digit observed is not the result of rounding, in the true sense of the word, but rather to observational limitations. It is possible to have a weight that has a value of 1.000050...0 grams but not likely. A value of 1.000050 can be uncertain in the last place due to measurement uncertainty and also to rounding. The value for the kilogram (the world's standard of mass) residing in the International Bureau in Paris is 1.000...0 kg by definition; all other mass standards will have an uncertainty for their assigned value.

VARIABILITY

Variability is inevitable in a measurement process. The operation of a measurement process does not produce one number but a variety of numbers. Each time it is applied to a measurement situation it can be expected to produce a slightly different number or sets of numbers. The means of sets of numbers will differ among themselves, but to a lesser degree than the individual values.

One must distinguish between natural variability and instability. Gross instability can arise from many sources, including lack of control of the process [1]. Failure to control steps that introduce bias also can introduce variability. Thus, any variability in calibration, done to minimize bias, can produce variability of measured values.

A good measurement process results from a conscious effort to control sources of bias and variability. By diligent and systematic effort, measurement processes have been known to improve dramatically. Conversely, negligence and only sporadic attention to detail can lead to deterioration of precision and accuracy. Measurement must entail practical considerations, with the result that precision and accuracy that is merely "good enough", due to cost-benefit considerations, is all that can be obtained, in all but rare cases. The advancement of the state-of-the-art of chemical analysis provides better precision and accuracy and the related performance characteristics of selectivity, sensitivity, and detection [1].

The inevitability of variability complicates the evaluation and use of data. It must be recognized that many uses require data quality that may be difficult to achieve. There are minimum quality standards required for every measurement situation (sometimes called data quality objectives). These standards should be established in advance and both the producer and the user must be able to determine whether they have been met. The only way that this can be accomplished is to attain statistical control of the measurement process [1] and to apply valid statistical procedures in the analysis of the data.

POPULATIONS AND SAMPLES

In considering measurement data, one must be familiar with the concepts and distinguish between (1) a population and (2) a sample. Population means all of an object, material, or area, for example, that is under investigation or whose properties need to be determined. Sample means a portion of a population. Unless the population is simple and small, it may not be possible to examine it in its entirety. In that case, measurements are often made on samples believed to be representative of the population of interest.

Measurement data can be variable due to variability of the population and to all aspects of the process of obtaining a sample from it. Biases can result for the same reasons, as well. Both kinds of sample-related uncertainty – variability and bias – can be present in measurement data in addition to the uncertainty of the measurement process itself. Each kind of uncertainty must be treated somewhat differently (see Chapter 5), but this treatment may not be possible unless a proper statistical design is used for the measurement program. In fact, a poorly designed (or missing) measurement program could make the logical interpretation of data practically impossible.

IMPORTANCE OF RELIABILITY

The term reliability is used here to indicate quality that can be documented, evaluated, and believed. If any one of these factors is deficient in the case of any data, the reliability and hence the confidence that can be placed in any decisions based on the data is diminished.

Reliability considerations are important in practically every data situation but they are especially important when data compilations are made and when data produced by several sources must be used together. The latter situation gives rise to the concept of data compatibility which is becoming a prime requirement for environmental data [1,2]. Data compatibility is a complex concept, involving both statistical quality specification and adequacy of all components of the measurement system, including the model, the measurement plan, calibration, sampling, and the quality assurance procedures that are followed [1].

A key procedure for assuring reliability of measurement data is peer review of all aspects of the system. No one person can possibly think of everything that could cause measurement problems in the complex situations so often encountered. Peer review in the planning stage will broaden the base of planning and minimize problems in most cases. In large measurement programs, critical review at various stages can verify control or identify incipient problems.

Choosing appropriate reviewers is an important aspect of the operation of a measurement program. Good reviewers must have both detailed and general knowledge of the subject matter in which their services are utilized. Too many reviewers misunderstand their function and look too closely at the details while ignoring the generalities. Unless specifically named for that purpose, editorial matters should be deferred to those with redactive expertise. This is not to say that glaring editorial trespasses should be ignored, but rather the technical aspects of review should be given the highest priority.

The ethical problems of peer review have come into focus in recent months. Reviews should be conducted with the highest standards of objectivity. Moreover, reviewers should consider the subject matter reviewed as privileged information. Conflicts of interest can arise as the current work of a reviewer parallels too closely that of the subject under review. Under such circumstances, it may be best to abstain.

In small projects or tasks, supervisory control is a parallel activity to peer review. Peer review of the data and the conclusions drawn from it can increase the reliability of programs and should be done. Supervisory control on the release of data is necessary for reliable individual measurement results. Statistics and statistically based judgments are key features of reviews of all kinds and at all levels.

METROLOGY

The science of measurement is called metrology and it is fast becoming a recognized field in itself. Special branches of metrology include engineering metrology, physical metrology, chemical metrology, and biometrology. Those learned in and practitioners of metrology may be called metrologists and even by the name of their specialization. Thus, it is becoming common to hear of physical metrologists. Most analytical chemists prefer to be so called but they also may be called chemical metrologists. The distinguishing feature of all metrologists is their pursuit of excellence in measurement as a profession.

Metrologists do research to advance the science of measurement in various ways. They develop measurement systems, evaluate their performance, and validate their

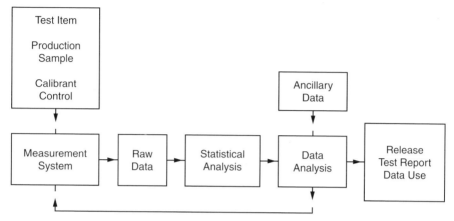

Figure 1.1. Role of statistics in metrology.

applicability to various special situations. Metrologists develop measurement plans that are cost effective, including ways to evaluate and assess data quality.

Statistics play a major role in all aspects of metrology since metrologists must contend with and understand variability.

The role of statistics is especially important in practical measurement situations as indicated in Figure 1.1. The figure indicates the central place of statistical analysis in data analysis which is or should be a requirement for release of data in every laboratory. When the right kinds of control data are obtained, its statistical analysis can be used to monitor the performance of the measurement system as indicated by the feedback loop in the figure. Statistical techniques provide the basis for design of measurement programs including the number of samples, the calibration procedures and the frequency of their application, and the frequency of control sample measurement. All of this is discussed in books on quality assurance such as that of the present author [1].

COMPUTER ASSISTED STATISTICAL ANALYSES

It should be clear from the above discussion that an understanding and working facility with statistical techniques is virtually a necessity for the modern metrologist. Modern computers can lessen the labor of utilizing statistics but a sound understanding of principles is necessary for their rational application. When modern computers are available they should be used, by all means. Furthermore, when data are accumulated in a rapid manner, computer assisted data analysis may be the only feasible way to achieve real-time evaluation of the performance of a measurement system and to analyze data outputs.

Part of the process involved in computer assisted data analysis is selecting a software package to be used. Many types of statistical software are available, with capabilities ranging from basic statistics to advanced macro programming features. The examples in the forthcoming chapters highlight MINITAB^{TM} [3] statistical software for calculations. MINITAB has been selected for its ease of use and wide variety of analyses available, making it highly suitable for metrologists.

The principles discussed in the ensuing chapters and the computer techniques described should be helpful to both the casual as well as the constant user of statistical techniques.

EXERCISES

1-1. Discuss the hierarchy: Data $\longrightarrow$ information $\longrightarrow$ knowledge.

1-2. Compare "hard" and "soft" data.

1-3. What are the similarities and differences of natural and experimental data?

1-4. Discuss discrete, continuous, and enumerative data, giving examples.

1-5. Why is an understanding of variability essential to the scientist, the data user, and the general public?

1-6. Discuss the function of peer review in the production of reliable data and in its evaluation.

REFERENCES

[1]   Taylor, J.K. *Quality Assurance of Chemical Measurements,* (Chelsea, MI: Lewis Publishers, 1987).

[2] Stanley, T.W., and S.S. Verner. "The U.S. Environmental Protection Agency's Quality Assurance Program," in *Quality Assurance of Environmental Measurements,* ASTM STP 967, J.K. Taylor and T.W. Stanley, Eds., (Philadelphia: ASTM, 1985), p. 12.

[3] *Meet MINITAB,* Release 14 for Windows (Minitab Inc. 2003).

CHAPTER 2

Obtaining Meaningful Data

Scientific data ordinarily do not occur out of the blue. Rather, they result from hard work and often from considerable expenditure of time and money. It often costs as much to produce poor quality data as to obtain reliable data and may even cost more in the long run. This chapter discusses some of the considerations that should be made and steps that should be taken to assure that data will be reliable and satisfactory for its intended purpose.

DATA PRODUCTION MUST BE PLANNED

The complexity of modern measurements virtually requires that a considerable amount of planning is needed to ensure that the data are meaningful [1]. While not the thrust of the present book, it can be said with a good degree of confidence that data quality is often proportional to the quality of advance planning associated with it. Experimental planning is now generally recognized as an emerging scientific discipline. This is not to say that scientific investigations up to recent times have not been planned. However, increased emphasis is being given to this aspect of investigation and a new discipline of chemometrics has emerged.

It is almost useless to apply statistical techniques to poorly planned data. This is especially true when small sets of data are involved. In fact, the smaller the data set, the better must be the preplanning activity. Any gaps in a data base resulting from omissions or data rejection can weaken the conclusions and even make decisions impossible in some cases. In fact, even large apparent differences between a control sample and a test sample or between two test areas may not be distinguished, statistically, for very small samples, due to a poor statistical power of the test. This is discussed further in Chapter 9.

The general principles of statistical planning have been described in earlier books (see, for example, References 2 and 3). In fact, Reference 2 contains a considerable amount of information on experimental design. An excellent book by Deming [4] has appeared recently that describes the state of the art of experimental design and planning of the present time from a chemometrics point of view.

Table 2.1. **Items for Consideration in Defining a Problem for Investigation**

What is the desired outcome of the investigation?
What is the population of concern?
What are the parameters of concern?
What is already known about the problem (facts)?
What assumptions are needed to initiate the investigation?
What is the basic nature of the problem?
   Research
   Monitoring
   Conformance
What is the temporal nature of the problem?
   Long-range
   Short range
   One-time
What is the spatial nature of the problem?
   Global
   Limited area
   Local
What is the prior art?
Other related factors

The advice presented here is to look behind the numbers when statistically analyzing and interpreting data. Unfortunately, all data sets do not deserve peer status and statistical tests are not necessarily definitive when making selections from compilations or when using someone else's data. While grandiose planning is not necessary in many cases, almost every piece of numerical information should be documented as to the circumstances related to its generation. Something akin to a data pedigree, i.e., its traceability, should be required.

The following sections in this chapter are included to call attention to the need for a greater concern for the data production process and to point out some of the benchmarks to look for when evaluating data quality.

THE EXPERIMENTAL METHOD

A proper experimental study consists in utilizing an appropriate measurement process to obtain reliable data on relevant samples in a planned measurement program designed to answer questions related to a well-defined problem. The identification and delineation of the problem to be investigated is a critical first step. Too often this important item is taken too lightly and even taken for granted. In the zeal to solve a problem or as a result of exigency, a program may be initiated with less than a full understanding of what the problem really is. Table 2.1 contains a listing of items to be considered in delineating a problem proposed for investigation. One can hardly devote too much effort to this most important first step.

In a classical book, E. Bright Wilson [5] describes the important steps in designing an experimental program. He cautions that the scope of work should be limited to something that can be accomplished with reasonable assurance. Judgment must be exercised to select the most appropriate parts for study. This will be followed by a statement of hypotheses that are to be tested experimentally. A successful hypothesis should not only fit the facts of the present case but it should be compatible with everything already known.

Care should be exercised to eliminate bias in experimentation. There is a danger of selecting only facts that fit a proposed hypothesis. While every possible variation of a theme cannot be tested, anything that could be critically related needs to be experimentally evaluated. Randomization of selection of samples and the order of their measurement can minimize bias from these important possible sources of distortion.

The experimental plan, already referred to in the previous section, is all important and its development merits all the attention that can be devoted to it. Its execution should be faithfully followed. A system of check tests to verify conformance with critical aspects of the plan is advisable. Finally, data analysis should incorporate sound statistical principles. Any limitations on the conclusions resulting from statistical and experimental deficiencies should be stated clearly and explicitly.

In an interesting article, entitled "Thirteen Ways to Louse Up an Experiment", C.D. Hendrix [6] gives the following advice:

- Decide what you need to find out or demonstrate.
- Estimate the amount of data required.
- Anticipate what the resulting data will look like.
- Anticipate what you will do with the finished data.

The rest of the article gives a lot of good advice on how to plan meaningful experimental programs and merits the attention of the serious experimenter.

What Data are Needed

The kind of data that are needed will be determined by the model of the problem investigated. This is discussed further under the heading, representativeness, in the section Data Quality Indicators. The selection of the species to be identified and/or quantified is a key issue in many chemical investigations. This is illustrated by investigations concerned with organic chemicals in that more than a million chemical compounds could be candidates for determination. Whether total organic substances, classes of organics, or individual compounds are to be sought could elicit differing opinions that may need to be resolved before measurements can begin. Unless there is agreement on what is to be measured, how can there be any agreement on the meaning of results of measurement?

Inorganic investigations have historically dealt with elemental analysis, which is to say total measurable elements. Many modern problems require further inorganic chemical information on such matters as the specific compounds that may be present, the biological availability of toxic substances, and the nature and spatial location of impurities in relatively pure substrates.

In both organic and inorganic analysis, it may be easier to specify what is needed than to experimentally evaluate their parameters. Data analysts need to be sure that the measurement process actually accomplishes what was desired of it. In light of modern requirements, much earlier data may need to be discarded (as painful as this might be) because of questions of what was measured as well as how well the measurements were done.

Amount of Data

The amount of data required to answer a question or to make a decision about it will depend on both the nature of the problem under investigation and the capability of the measurement program to provide data of adequate quality.

Knowing the expected variability of the measurement process and of the samples to be investigated, one can estimate the number of samples and measurements required to attain a desired level of precision. Statistical techniques applicable to such estimations are described in later chapters. Further guidance on these matters is provided in the author's book on quality assurance of measurements [7]. As small differences in several populations are of concern, these questions become of critical importance. It is obvious that cost-benefit considerations become important in designing such measurement programs. It is futile to conduct an experimental investigation in such areas unless adequate resources are made available to support the measurement program that is required.

Quality Considerations

Much is being said these days about data quality and what is needed to assure that it meets the needs of decision processes. The following sections briefly review the concept of data quality and identify the characteristics that may be used to specify quality in advance and to evaluate the final product.

DATA QUALITY INDICATORS

Data consist of numerical values assigned to some characteristic of a population under study. The naming of the characteristic may seem to be a trivial exercise and indeed this is so in many cases. However, the **qualitative identification** of what is measured must be known with confidence approaching certainty if the data are to have any use whatsoever [7].

The qualitative identification can pose problems as the limits of measurement and detection are neared. Most chemical methodology suffers some degree of non-selectivity and problems can arise when investigations of possible interferents are done inadequately. Problems related to speciation are also possible.

In organic chemistry, this can concern isomers, misidentified compounds, and problems of resolution of measuring apparatus. In inorganic analysis, elemental analysis has been almost the sole objective, up to recent times, with little regard to oxidation states and almost no consideration of what compounds were actually present. Questions of complexation in the natural environment largely have been ignored so that total element may have little relation to available element in many cases. All this has changed in recent years and such questions increasingly must be answered in addition to simply finding the quantitative amounts of what may be present.

In summary, modern science and technology are making new demands on the qualitative identification of the parameter measured and/or reported that require careful consideration of what was measured as well as its numerical aspects.

The **quantitative accuracy** of what is measured is an obvious indicator of data quality. Because of inescapable variability, data will always have some degree of uncertainty. When measurement plans are properly made and adequately executed, it is possible to assign quantitative limits of uncertainty to measured values. The statistical techniques used for such assignment as well as those used to make decisions, taking into account well-documented uncertainty, constitute the bulk of the remainder of the content of this book.

Three additional indicators of data quality will be described briefly. The **representativeness** is a prime consideration when using data. This term describes the degree to which the data accurately and precisely represent a characteristic of a population parameter, variation of a property, a process characteristic, or an operational condition. It is difficult to quantify representativeness, yet its importance is obvious. Professional knowledge and opinion during planning enhance the chances of obtaining representative data while expert judgment must be exercised when deciding how representative acquired data really are.

Completeness is a measure of the amount of data obtained as compared with what was expected. Incomplete data sets complicate their statistical analysis. When key data are missing, the decision process may be compromised or thwarted. While the percentage of completeness of data collection can be ascertained in most cases, questions of the serious consequences of critical omissions is a matter for professional judgment.

Comparability of data from various sources is a requirement for combination and intercomparisons. It is achieved by proper design of measurement programs and by demonstrated peer performance of participants. Statistics can aid when deciding whether peer performance has been achieved and provides the basis for numerical merging of data sets. However, representativeness also comes into consideration since numerical merging of unlike data is irrational. The statistician must always be aware of this problem but may have to depend on subject area experts for advice on comparability from the representativeness point of view.

DATA QUALITY OBJECTIVES

Data quality objectives (DQOs) consist of quantitative specifications for the minimum quality of data that will permit its use in a specific investigation. They must be realistic with respect to what is needed and what it is possible to achieve. Cost and benefit considerations will be involved in most cases. Statements of what can be achieved should be based on sound evaluation of the performance capability of methodology and of laboratories.

All of the data quality indicators named above are useful and should be addressed when specifying DQOs. DQOs developed in advance do not guarantee data of adequate quality but their absence can lead to false expectations, and data of inadequate quality due to failure to appreciate what is needed. Qualification of laboratories on the basis of their ability to achieve DQOs is necessary and such a process depends heavily on statistical evaluation of their performance on evaluation samples.

SYSTEMATIC MEASUREMENT

It is becoming clear that the production of data of known and adequate quality depends on systematic measurement [7]. The methodology used must be selected to meet the DQOs, calibration must be systematized, and a quality assurance program must be followed. Samples measured must have a high degree of relevancy to the problem investigated and all aspects of sampling must be well planned and executed. All of these aspects of measurement must be integrated and coordinated into a measurement system.

Measurements made by less than a well-designed and functioning measurement system are hardly worthy of serious statistical analysis. *Statistics cannot enhance poor data.*

QUALITY ASSURANCE

The term quality assurance describes a system of activities whose purpose is to provide evidence to the producer or user of a product or a service that it meets defined standards of quality with a stated level of confidence [7]. Quality assurance consists of two related but separate activities. *Quality control* describes the activities and procedures utilized to produce consistent and reliable data. *Quality assessment* describes the activities and procedures used to evaluate that quality of the data that are produced.

Quality assurance relies heavily on the statistical techniques described in later chapters. Quality control is instrumental in establishing statistical control of a measurement process. This vital aspect of modern measurement denotes the situation in which a measurement process is stabilized as evidenced by the ability to attain a limiting mean and a stable variance of individual values distributed about it. Without statistical control one cannot believe logically that a measurement process is measuring anything at all [8].

While of utmost importance, the attainment of statistical control cannot be proved, unequivocally. Rather one has to look for violations such as instability, drifts, and similar malfunctions and this should be a continuing activity in every measurement laboratory. Provided a diligent search is made, using techniques with sufficient statistical power, one can assume the attainment of statistical control, based on the lack of evidence of noncontrol.

Quality assessment provides assurance that statistical control has been achieved: quality assessment checks on quality control. Replicate measurements are the only way to evaluate precision of a measurement process, while the measurement of reference materials is the key technique for evaluation of accuracy. Utilization of the statistical techniques described later, in conjunction with control charts, is essential to making decisions about measurement system performance.

IMPORTANCE OF PEER REVIEW

Peer review is an important and ofttimes essential component of several aspects of reliable measurement. Participants in measurement programs need to review plans for adequacy and attainability. Subject matter experts provide review to see that the right data are taken and that the results can be expected to provide definitive decisions on the issues addressed. Statisticians are needed to review the plans of nonstatisticians and even of other statisticians from the point of view of statistical reliability and appropriateness. Unless essentially faultless plans are followed that achieve consensus approval, the final outcome of a measurement program can hardly hope to gain acceptance.

Review of the data analysis is likewise required. Reports must withstand critical review and the conclusions must be justified on both technical and statistical grounds. Reporting should be consistent with current practice and with the formats of related work if they are to gain maximum usefulness.

EXERCISES

2-1. Discuss the concept of "completeness" as an indicator of data quality.

2-2. Discuss the concept of "representativeness" as an indicator of data quality.

2-3. Discuss the concept of "comparability" as an indicator of data quality.

2-4. What is meant by data quality objectives and why are they of great importance in the assurance of data quality?

2-5. What is meant by statistical control of a measurement process?

2-6. Define quality assurance and discuss its relation to data quality.

REFERENCES

[1] Taylor, J.K., "Planning for Quality Data," *Mar. Chem.* 22: 109-115 (1987).

[2] Natrella, M.G., "Experimental Statistics", NBS Handbook 91, National Institute of Standards and Technology, Gaithersburg, MD 20899. Note: This classical book has been reprinted by Wiley-Interscience to facilitate world-wide distribution and is available under the same title (ISBN 0-471-79999-8).

[3] Youden, W.J., *Statistical Methods for Chemists,* (New York: John Wiley & Sons, 1951).

[4] Deming, S.N., *Experimental Design; A Chemometrics Approach*, (Amsterdam: Elsevier, 1987).

[5] Wilson, E.B., *An Introduction to Scientific Investigation* (New York: McGraw-Hill Book Company, 1952).

[6]   Hendrix, C.D., "Thirteen Ways to Louse Up an Experiment," *CHEMTECH*, April (1986).

[7]   Taylor, J.K., *Quality Assurance of Chemical Measurements,* (Chelsea, MI: Lewis Publishers, 1987).

[8]   Eisenhart, C., "Realistic Evaluation of the Precision and Accuracy of Instrument Calibration Systems," in *Precision Measurement and Calibration: Statistical Concepts and Procedures,* NBS Special Publication 300. Vol. 1, (Gaithersburg, MD: National Institute of Standards and Technology).

CHAPTER 3

General Principles

Statistics is looked upon by many scientists and engineers as an important and necessary tool for the interpretation of their measurement results. However, many have not taken the time to thoroughly understand the basic principles upon which the science and practice of statistics are based. This chapter attempts to explain these principles and provide a practical understanding of how they are related to data interpretation and analysis.

INTRODUCTION

Everyone who makes measurements or uses measurement data needs to have a good comprehension of the science of statistics. Statistics find various uses in the field of measurement. They provide guidance on the number of measurements that should be made to obtain a desired level of confidence in data, and on the number of samples that should be measured whenever sample variability is of concern. They especially help to understand the quality of data. Nothing is ever perfect and this is very true of data. There is always some degree of uncertainty about even the most carefully measured values with the result that every decision based on data has some probability of being right and also a probability of being wrong. Statistics provide the only reliable means of making probability statements about data and hence about the probable correctness of any decisions made from its interpretation.

From what was said above, it may be concluded that statistics provide tools, and indeed very powerful tools, for use in decision processes. However, it should be remembered that statistical techniques are only tools and should be used for enlightened guidance and certainly not for blind direction when making decisions. A good rule to follow is that if there is conflict between intuitive and statistically guided conclusions, one should stop and take careful consideration. Was one's intuition wrong or were wrong statistical tools used?

> Statistical Techniques
> are
> TOOLS
> Rather Than
> ENDS

Statistics of one kind or another find important uses in everyday life. They are used widely to condense, describe, and evaluate data. Large bodies of data can be utterly confusing and almost incomprehensible. Simple statistics can provide a meaningful summary. Indeed, when a mean and some measure of the spread of the data are given, one can essentially visualize what an entire body of data looks like.

> USES for STATISTICS
> - Condense Data
> - Describe Data
> - Assist in Making Decisions

The thrust of this book is to show how to use statistics effectively for the evaluation of data. It should be remembered that statistics is a scientific discipline in itself. There are many valuable ways that statistics can be used that are too complicated to be discussed in a simple presentation and must be left to the professional statistician. However, every scientist needs to understand basic statistical principles for guidance in effective measurement and experimentation, and there are many things that one can do for one's self. Even if nothing else is gained, this book should help to engender a better dialogue when seeking the advice or assistance of a statistician, and to promote better understanding in designing and implementing measurement programs.

KINDS OF STATISTICS

There are basically two kinds of statistics. **Descriptive statistics** are encountered daily and used to provide information about such matters as the batting averages of baseball players, the results of public opinion polls, rainfall and weather phenomena, and the performance of the stock market. However, the statistics of concern here are **inductive statistics**, based on the description of well-defined so-called populations, that may be used to evaluate and make predictions and decisions based on measurement data.

DECISIONS

Numerous decisions are made every day, often based on the interpretation of

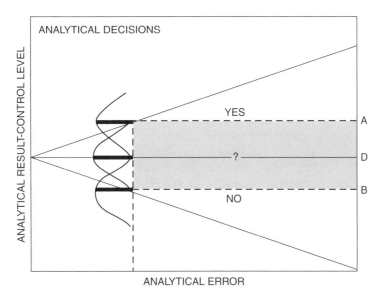

Figure 3.1. Measurement decision.

measurement data. In fact, decision making is the ultimate use of most of the results of measurement. A simple example of such decisions is shown in Figure 3.1. There may be a need to decide whether the property of a material exceeds some critical level, D. If so, the answer is **YES**: if not, the answer will be **NO**. If the measured value is well up into the YES area, or well down in the NO area, the decision is easy to make. When it is exactly at D, it is puzzling since the slightest amount of measurement error would make the true value higher or even lower than D. In fact, even when a measured value such as A is obtained which is apparently greater than D, the same dilemma is present, and similarly in the case of B. The bell-shaped curves indicate the probable limits for the relation of the measured value to the true value. The way these limits are calculated and used in the decision process will be discussed in Chapter 5.

It should be clear from the above figure that the shaded area is the area of indecision for the data. It has to be reasonably small to make data useful. Above all, the limits must be known in order to make any conclusions, whatsoever, about the measured values. The width of the crosshatched area depends on the numerical value of the standard deviation and the number of independent measurements that are made.

ERROR AND UNCERTAINTY

The word statistics implies data and the data have to represent the facts.

The Word Statistics Implies Data
• Data Should Reflect Facts • No Matter How Carefully Incorrect Data are Analyzed, the Results are Meaningless

If the data do not represent the facts, no conclusions will be meaningful. Data can be erroneous due to any of a number of causes, ranging from faulty measurement practices to miscalculations, and such data should not be included in a data set. Several techniques that may be used to statistically decide whether suspected data are outliers or whether they are fringe area data will be discussed in Chapter 6. No matter how carefully erroneous data are statistically treated, the results will not be meaningful. One should look critically at data at all times not only to eliminate erroneous items but to help to identify malfunction of measurement processes whenever such should occur.

Data that are error-free can and will be uncertain, due to the natural variability of the measurement process. This is one of the main reasons why statistics are used to analyze and interpret data, and will be discussed later. Statistics provide a means to estimate the uncertainty of data and to decide whether apparent differences are significant.

KINDS OF DATA

Referring to Figure 3.2, there are basically three kinds of data. The first (Figure 3.2A) consists of single measurements on a number of items. The second (Figure 3.2B) consists of a number of measurements on the same object. The third (Figure 3.2C) is a combination of the above in which several measurements are made on each of several objects. Each kind requires a somewhat different statistical treatment.

ACCURACY, PRECISION, AND BIAS

A review of the concepts of accuracy, precision, and bias would seem to be appropriate at this point. Accuracy refers to the closeness of the measured value to the true value of what is measured. Actually, accuracy is never evaluated but rather inaccuracy, which is to say the departure of the measured value from the true value. Any departure is of course an error and one should want to know the reason for it.

There are three sources of error in measurements and they can be classed as **systematic**, **random**, and just plain **blunders**, or **mistakes**.

Each systematic error associated with a given measurement process is always of the same sign and magnitude. It persists measurement after measurement. When its existence is established, such an error is called a bias, and reasonable effort should be made to correct for it. Sometimes the observed bias is the result of the concurrence

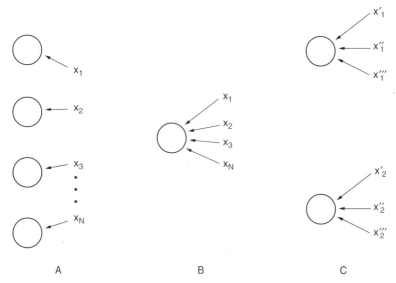

A   B   C

Figure 3.2. Types of data.

of several biases that cannot or at least have not been individually identified. One of the purposes of statistical treatment of data is to decide whether an apparently erroneous result is real and indicates a bias or whether it could happen as the result of chance variability, even in a well-behaved measurement system. There can be, of course, biases that have not been identified as such. Also, there are limits to how well one can correct for known biases, and this inadequacy must be considered when limits of uncertainty are assigned to data.

The second type of uncertainty results from random causes that produce fluctuations in both sign and magnitude, the latter within well-defined limits, however. In the long run, the random error averages out to zero. The random error accounts for the variability of individual measurements and it will be shown that it can be statistically characterized by what is called a standard deviation. This term is thus a measure of the dispersion of the data around a mean or average value. When the value of the standard deviation is small, the data cluster closely around the mean; when it is large, the spread is greater.

The third kind of uncertainty may be called blunders. They happen when mistakes are made, knowingly or unknowingly. Errors resulting from them are not statistically manageable and, in fact, they can invalidate an otherwise good set of data. A measurement system that is unstable and fraught with blunders is not in statistical control and cannot be relied upon to produce useful data.

Bias in a measurement system describes the situation where the limiting mean (the value approached by the average based on very many measurements) differs from the true value. The upper part of Figure 3.3 contains three curves that represent the distribution of measured values around limiting means, resulting from three different

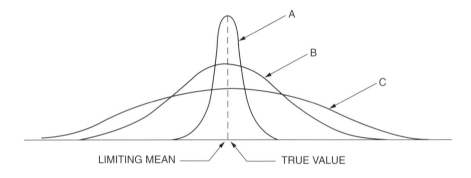

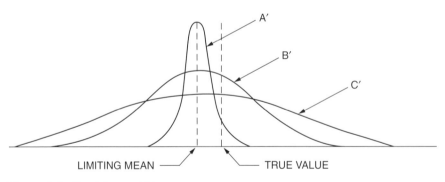

Figure 3.3. Precision and bias.

measurement processes. All of the processes are unbiased because their limiting means coincide with the true value. However, they differ in their precisions as indicated by the spread of the curves. Thus, A is the most precise and C is the least precise. The lower half of the figure represents the results of three other processes, all of which are biased because the limiting means do not coincide with the true value. Again, the processes differ in their precisions.

Because any value other than the true value is erroneous and hence inaccurate, it is clear that inaccuracy of an individual measured value results from two sources, precision (really imprecision) and bias. From the lower part of Figure 3.3, it is clear that a bias smaller than the dispersion of a set of data will be difficult to detect unless conclusions are based on the mean of a sufficiently large number of measurements.

Accordingly, the precision must be smaller than any unacceptable bias in a reported value in order for it to be useful. After, and only after, such a requirement is attained can one look for bias in a measurement process and its data outputs.

The third kind of uncertainty in data results from blunders. While blunders will not be dealt with here, measurement experts and data users must be ever mindful of their possible existence and take effective practices to screen data sets for their existence. Often, they produce outliers, a subject that will be addressed in Chapter 6.

Before leaving this subject, it should be remembered that the words precise and accurate are commonly used as general descriptors of measurement systems and even data sets. Such usage is always relative to a perceived data requirement. What would be very precise for one use could be imprecise in another situation, and likewise for accuracy. Also, it should be remembered that an unbiased measurement system will produce inaccurate results due to precision considerations. Thus, the evaluation of both precision and bias are always of concern whenever accuracy is considered.

STATISTICAL CONTROL

Dr. Churchill Eisenhart of the National Bureau of Standards, (now known as the National Institute of Standards and Technology) one of the most eminent statisticians of our time, once stated [1]:

Until a measurement system is in a state of statistical control, it cannot be believed in any logical sense that it is measuring anything at all.

What did he mean by this? A measurement process must exist and it must be stable. Although each individual measurement can be expected to differ somewhat from each other, the pattern of deviations must be stable (i.e., a constant dispersion) and the averages of a large number of measurements should zero in on a particular value, called the limiting mean. The individual values will of course be randomly distributed around this limiting mean, if a large number of measurements are considered. Only in a state of statistical control is a statistical treatment of the data meaningful. Otherwise it is a futile exercise in mathematics.

Data Descriptors

It is a common observation that the individuals in any sizable body of data tend to cluster around some central value. This value is often quoted as a descriptor of some general characteristic of the entire set. The single data descriptor most commonly used is the average value or arithmetic mean. However, this has its limitations as will be seen in the following sets.

Set A   10, 9, 11, 10, 9, 9, 11, 10, 10, 11
Avg = 10.0

Set B   7, 12, 9, 8, 15, 12, 6, 10, 8, 13
Avg = 10.0

Set C   14, 13, 15, 14, 13, 6, 7, 6, 6, 6
Avg = 10.0

All of the data sets in the box above have the same average value, $\overline{X}$, but they differ widely. The values in set A range from 9 to 11, while those in set B range from 6 to 15. Set C differs further in that it appears to contain two sets, one ranging from 6 to 7 and the other from 13 to 15. Clearly, something in addition to the average is needed to describe a data set. The range is one of the useful descriptors, i.e., the difference between the largest and the smallest value in a set. As more data are added to a set, the range can get larger, but never smaller than some initially estimated value. This is one limitation on the use of the range as a general descriptor of the variability of data.

The arithmetic mean is the descriptor first thought of when the word mean is mentioned. However, the geometric mean is sometimes more appropriate. It is calculated as the nth root of the product of n data points, or equivalently by the use of logarithms as shown in the box below.

Arithmetic Mean $$\overline{X} = \frac{X_1 + X_2 + ... + X_n}{n}$$

Geometric Mean $$(X_1 \times X_2 \times ... \times X_n)^{1/n}$$

or

$$\text{Antilog of } [(\log X_1 + \log X_2 + ... + \log X_n)/n]$$

Median   Middle Value

Mode   Most Frequent Value(s)

The mean is not always the most appropriate descriptor of the "center" of a data set. For some purposes, the median or middle value may be better, especially in small data sets. The median is obtained by first lining up the individual data points in the order of their numerical value. In the case of an odd number of points, the middle value is the median. In the case of an even number of points, the average of the two points straddling the center is the median.

Example In the set

6, 7, 8, 9, 10, 12, 12, 13, 14, 15, 17, 19, 20

the median value is 12. In the case of the data set

6, 7, 8, 9, 10, 12, 12, 13, 14, 15, 17, 19, 20, 23

the median value is 12.5.

Another descriptor is the mode which is the value occurring most frequently in the data set, usually but not always located somewhere near the center. Thus in both sets above, the mode is 12. There can be several modes in a set of data, resulting from the fact that they represent several populations. For example, without going into details, measurements of particle size distributions in the atmosphere can exhibit two modes due to particles arising from combustion processes (usually fine particles) and those from abrasion processes (usually resulting in coarser particles). If all of such data were lumped together, one could reach an erroneous conclusion about the average size of particles in the atmosphere. The situation is even more complicated in some cases when number averages and weight averages may need to be considered, obviously beyond our present scope. However, one must always be aware of such possible complications when statistically analyzing data sets.

The subject of the central value for a data set is discussed further in Chapter 9.

DISTRIBUTIONS

It should be clear by now that the way data are distributed around a central value is important, scientifically. However, it is just as important from a statistical point of view, because it will affect how one should statistically treat one's data. While the values of individual data points for objects or samples of material may vary according to a number of patterns, measurement data often follow a rather simple distribution. The data points are typically distributed symmetrically about the mean. Small deviations from the average usually occur more often than larger differences and very large differences occur only rarely. When a frequency distribution or histogram is constructed of large numbers of measurement results, one often obtains the familiar bell-shaped curve which has been mathematically described by the mathematician Gauss and is called the Gaussian or normal distribution.

The mathematical equation for such a distribution is as follows.

$$y = \frac{1}{\sigma\sqrt{2\pi}} e^{-(X-\mu)^2/2\sigma^2}$$

In this equation, μ represents the population arithmetic average or mean value and σ represents the root mean square deviation of the individuals from the average. It is also the point of inflection (point of change of curvature) of the curve. σ is called the population standard deviation, so named because it is the deviation that statisticians have adopted as the standard for characterizing the dispersion of a normal population. Two thirds of the population will be contained within the limits of ±1 σ about the mean. Figure 3.4 illustrates the percentage of the population covered by other multiples of σ. Also, this matter will be discussed further when considering statistical tolerance intervals in Chapter 5.

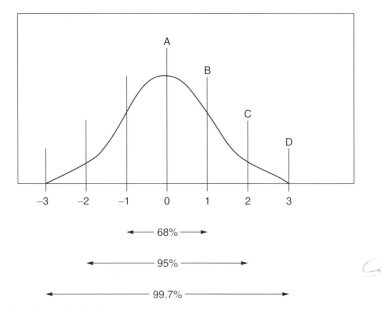

Figure 3.4. Normal distribution.

As already stated, various kinds of distributions can be observed when various phenomena are considered. Several of these are shown in Figure 3.5. The normal distribution (Figure 3.5A) so called because it is "normally" expected (other distributions are not abnormal), has already been described. The uniform distribution (Figure 3.5B) is one in which all possibly different individuals occur with the same frequency. Thus, in rolling a six-sided die, 1 to 6 should occur with the same frequency in a large number of trials, as shown in Figure 3.5B. Drawing cards from a well-shuffled deck is another example.

The log normal distribution (Figure 3.5C) fits much data when large differences from the central value occur more frequently than small departures, as in the case of environmental data, for example. The log normal distribution becomes a normal distribution when the logarithms of the data are used as if they were the actual data points. (See Chapter 4 for more about this.)

Variations of the normal distribution are shown in Figure 3.6. The bimodal distribution shown in Figure 3.6B has already been commented on. The multimodal distribution shown in Figure 3.6C results from a number of unresolved distributions lumped together unknowingly, because no effort was made to resolve them, or resolution may have been difficult if not impossible. Such distributions may occur more frequently than supposed.

Distribution shown in Figure 3.6D results when a data set contains both numerical data and attribute data such as "nondetect" data, for example. It is obvious that the statistics of such a data set will be influenced by the omission of this part of it. To

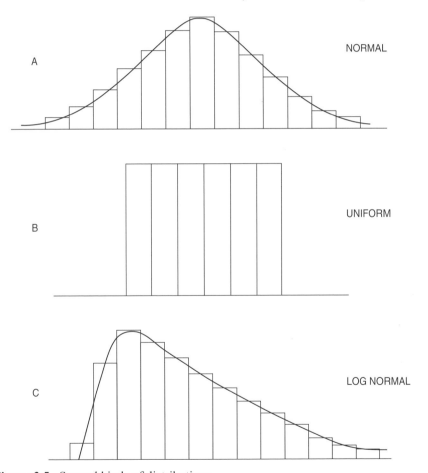

Figure 3.5. Several kinds of distributions.

minimize such a problem, the arrangement shown in Figure 3.6E is sometimes used. This is an example of a data set containing "nondetects" in which all of the values in the nondetect region are given the value of one half the minimum level of detection. The merits of such a procedure are outside of the scope of this book.

Figure 3.6F represents the situation when outliers are present. Again, the statistics of description of the population would be warped by including such extraneous data.

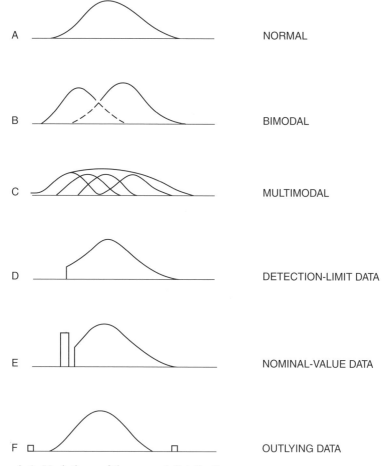

Figure 3.6. Variations of the normal distribution.

TESTS FOR NORMALITY

How can the distribution of a set of data be determined? If one had a very large amount of data, and especially all possible data, its histogram would be indicative, but would not be quantitatively interpretable. Figure 3.7 shows histograms for two sets of experimental data. Although both plots are based on a reasonably large number of points, a perfect normal distribution is not realized. Some of the groups are overpopulated while others contain less than what would be predicted for a normal distribution. It is left to the reader to decide whether there is a good approximation to normal distribution in either or both of these cases.

Another and perhaps easier way to decide on normality is to make a probability plot of the data set. This will work reasonably well if the sets are not too small, that is to say for more than 10 points, and ideally for a much larger number of points. The data points are arranged in ascending order of their magnitude, called ranking. One then makes the calculations indicated in the box shown on the following page.

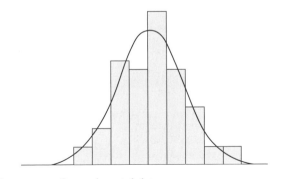

Figure 3.7. Histograms of experimental data.

PROBABILITY PLOTS

1. Rank data from smallest to largest value

        n=total number of data points
        i=rank

2. Calculate probability plot position

$$F_i = 100[(i - 0.5)/n]$$

3. Plot on appropriate probability paper

4. Interpret on basis of goodness of fit to a straight line

The data are then plotted with the value on the Y axis against the plot position, F_i, on the X axis. A special plotting paper is used. Reference 2 is given as one of a possible number of sources for probability plotting paper. To test for normal distribution, the Y axis is linearly subdivided; to test for a log normal distribution, the Y axis is logarithmically subdivided. In either case, the X axis is nonlinear with graduations bunched in the middle and increasing in each direction.

Example   Consider the data set

$$6.2, 7.4, 10.4, 7.9, 8.7, 8.6, 10.0, 8.1, 9.3, 9.0$$

on ranking

X	6.2,	7.4,	7.9,	8.1,	8.6,	8.7,	9.0,	9.3,	10.0,	10.3
i	1	2	3	4	5	6	7	8	9	10
F_i	5	15	25	35	45	55	65	75	85	95

Using the kind of graph paper described above, normally distributed data will fall on a straight line. The resulting plot for the data above is shown in Figure 3.8. One would conclude that the values are normally distributed.

Example   Consider the data set

6.2, 7.4, 10.4, 7.9, 8.7, 8.6, 10.0, 8.1, 9.3, 9.0

on ranking

X	6.2,	7.4,	7.9,	8.1,	8.6,	8.7,	9.0,	9.3,	10.0,	10.3
i	1	2	3	4	5	6	7	8	9	10
F_i	5	15	25	35	45	55	65	75	85	95

Using the kind of graph paper described above, normally distributed data will fall on a straight line. The resulting plot for the data above is shown in Figure 3.8. One would conclude that the values are normally distributed.

Example Consider the data set

7.0, 9.9, 12.5, 24.0, 70.0, 7.0, 16.0, 30.5, 19.0, 41.0

If these data are ranked, calculations are made, and plotted as before, what would be the conclusion? (It is suggested that the reader may want to do this for her- or himself.) Data like the above suggest a log normal distribution. This assumption could be tested by plotting on paper like that used before except that the Y axis is logarithmically subdivided. If the data fall on a straight line, the assumption of a log normal distribution would be justified. Figure 3.9 is a plot of the above data on logarithmically divided probability paper. If logarithmetic probability paper is not available, the logarithms of X, rather than X, could be plotted on regular arithmetic probability paper. Figure 3.10 is an example of such a plot.

Because of the natural variability of real world data, the points will not usually fall on a perfect straight line. When many data points are available, the deviations will balance out. For small sets of data, this may not occur and apparent curvature may be indicated. In this case the data analyst will need to decide whether the curvature is real or just an artifact of that particular set of data. This situation occurs quite frequently and that is why a reasonably large set of data is needed when making decisions about normality. Some of the kinds of plots that may be obtained are illustrated in Figure 3.11. There are statistical ways to decide on the significance of departures, depending on the number of data points plotted. This is discussed further in connection with "goodness of fit" in Chapter 9.

Two special kinds of departures from the normal distribution are skewness, in which the data are unsymmetrically distributed around a mean, and kurtosis (curvature), in which the data are abnormally compressed or are more spread out than for a true normal distribution. Guidance on whether an apparent skewness is real (and

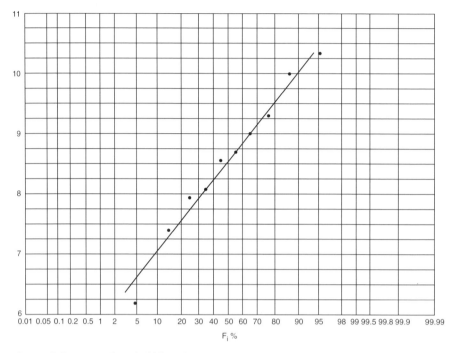

Figure 3.8. Normal probability plot.

therefore inconsistent with a normal distribution) can be obtained by calculating the skewness factor, g_1, using the equation given in Figure 3.12. Instructions for calculating σ are given in Chapter 4.

For a perfect normal distribution, g_1 will be zero. A negative value is due to skewness toward lower values while a positive value indicates excess higher values. For small data sets, one often gets values differing from zero. Statistical tests are available to help judge the significance of any observed departure. Table 3.1 contains the range within which g_1 is expected to lie, with 95% confidence, due to data variability considerations, for several sizes of samples. Thus for a set containing 10 points, g_1 would be expected to lie within the range of values, -0.95 to $+0.95$ with 95% confidence. For a set of 25 points the corresponding range would be expected to be -0.711 to $+0.711$. Transgression of these bounds would indicate a non-normal distribution at the probability level of the test.

The kurtosis factor, g_2, can be calculated according to the expression given in Figure 3.13. Again, zero is expected for a normal distribution. A negative value indicates sharper while a positive value indicates data that are more spread out than normal. The significance of a g_2 value differing from zero can be judged on a probability basis, as mentioned above. Table 3.2 contains values within which g_2 would be expected to lie with a given probability, due to variability considerations.

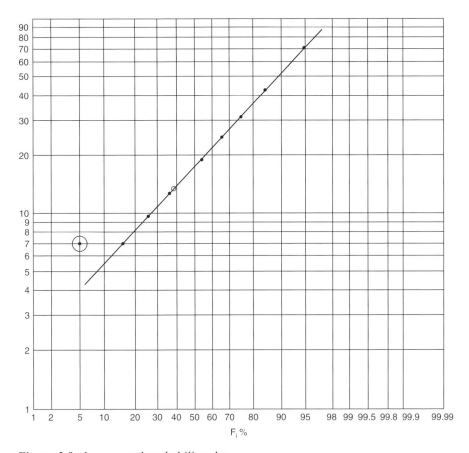

Figure 3.9. Log normal probability plot.

Before leaving the subject of the normal distribution, it should be pointed out that the means of sets of data that are not normally distributed often show normal distribution. This results from what is called the central limit theorem. As an example, consider the casting of a six-sided die. A uniform distribution of the single casts is expected and realized when a sufficient number of casts are made. Figure 3.14 shows the frequency distribution of 126 die casts.

One would expect to get 21 of each number, 1 to 6, and it is seen that this was approximately true. When the casts are broken up into groups that are averaged, one would expect to obtain an average value of 3.5. The data used for Figure 3.14 were separated into 16 consecutive groups of ten, averaged, and plotted on arithmetic normal probability paper. The plot is shown in Figure 3.15. A straight line is not a perfect but a reasonable fit of the data.

There are, of course, a number of other tests that could be used for judgment of normality. These will be found in books providing a more extensive treatment of statistics, to which the reader is referred.

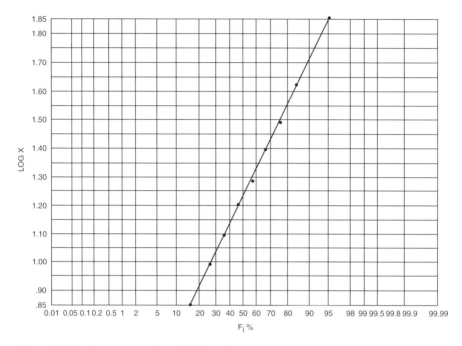

Figure 3.10. Log × normal probability plot.

BASIC REQUIREMENTS FOR STATISTICAL ANALYSIS VALIDITY

Fortunately, the results of many measurement processes may be considered to be normally distributed. However, data sets must meet the additional requirements listed in the box below before they may be analyzed by the statistical techniques that will be described later.

> BASIC REQUIREMENTS
> Stable
> Independent
> Random

First, the process generating the data must be stable. Second, every data point should be independent, that is to say uninfluenced by any other data point in the set. Third, the data points should be randomly distributed around the mean. There is no way to unequivocally prove that these requirements are met in any situation. The only thing that can be done is to look for departures from the requirements. This should always be done before applying statistical treatment to any data set. If no significant

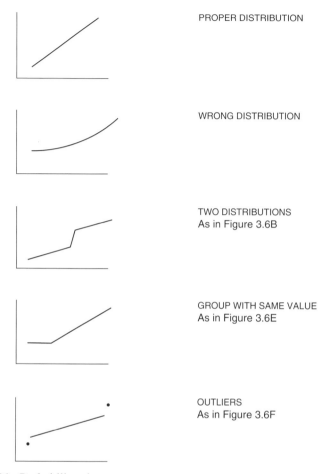

Figure 3.11. Probability plots.

violations are found, there is no reason to believe that there are violations (note that this is different from saying that the requirements *have* been met). The statistician calls this the test of the null hypothesis. In other words, it is the concept of "innocent until proven guilty".

Whenever graphical plotting of data is possible, for example according to the sequence or time of measurement, gross departures should be identifiable. Consider the typical plots shown in Figure 3.16. Instability is evidenced by trends, jumps, and shifts in data plots. Systematic patterns indicate a lack of randomness. The eye is a good instrument for discovering patterns or trends. In fact, it can be too good so that one may become unduly concerned about microdefects in a data set. It will be discussed later how statistical tests such as t-tests and F-tests (see Chapter 5) may be used to help judge the significance of apparent deviations.

$$g_1 = \frac{\sum_{i=1}^{n} (x_i - \bar{x})^3}{n\sigma^3}$$

$g_1 = -$ $g_1 = 0$ $g_1 = +$

Figure 3.12. Skewness.

Table 3.1. **Limits for the Skewness Factor, g_1, in the Case of a Normal Distribution**

Size of Sample	Probability Level	
(n)	5%	1%
5	−1.058 to +1.058	−1.342 to +1.342
10	−.950 to +.950	−1.397 to +1.397
15	−.862 to +.862	−1.275 to +1.275
20	−.777 to +.777	−1.152 to +1.152
25	−.711 to +.711	−1.061 to +1.061
30	−.661 to +.661	−.982 to +.982
35	−.621 to +.621	−.921 to +.921
40	−.587 to +.587	−.869 to +.869
45	−.558 to +.558	−.825 to +.825
50	−.533 to +.533	−.787 to +.787
100	−.389 to +.389	−.567 to +.567
200	−.280 to +.280	−.403 to +.403
1000	−.127 to +.127	−.180 to +.180
5000	−.057 to +.057	−.081 to +.081

Adapted from Table V B of NBS Technical Note [3]. Values adapted from *Bimetrika* 60: 172 (1973).

The requirement for independence is important since it determines the degrees of freedom (more fully discussed in Chapter 4) associated with statistical parameters. Correlations of the data with common factors need to be looked for (see Chapter 9 for more about this). Often some aspect of the measurement process can be common to a set of measurements, such as the same calibration, for example. In such a case, all of a set of measurements could be correlated with respect to the same calibration. The differences of successive measured values are correlated since a high reading for

$$g_2 = \frac{\sum\limits_{i=1}^{n}(x_i - \bar{x})^4}{n\sigma^4} - 3$$

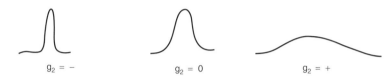

$g_2 = -$ $g_2 = 0$ $g_2 = +$

Figure 3.13. Kurtosis – g_2 (curvature).

Table 3.2. Limits for the Kurtosis Factor, g_2, in the Case of a Normal Distribution

Size of Sample	Probability Level	
n	5%	1%
200	−.49 to +.57	−.63 to +.98
400	−.36 to +.41	−.48 to +.67
600	−.30 to +.34	−.40 to +.54
800	−.26 to +.29	−.35 to +.46
1000	−.24 to +.26	−.32 to +.41

Excerpted from Table V C of NBS Technical Note 756 [3].

a preceding value results in a high difference for it and a low value for a succeeding difference (see Chapter 9 for a further discussion of this). Because correlations (especially unsuspected ones) can greatly influence and often limit conclusions, the experimenter should think about such matters whenever designing a measurement program or analyzing a set of data. Statisticians have clever ways to look for correlations. This is very important because hidden replications result in interdependence of measurements that may appear on cursory inspection to be independent. Techniques for testing nonrandomness are discussed in Chapter 9.

MINITAB

Introduction to MINITAB

Now that the reader is familiar with some basic methods for summarizing data, computer software serving the same purpose can be introduced. Chapter 1

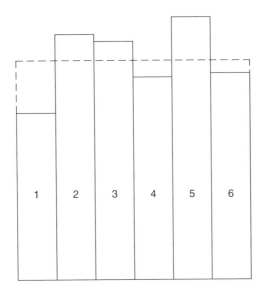

Figure 3.14. Experimental uniform distribution.

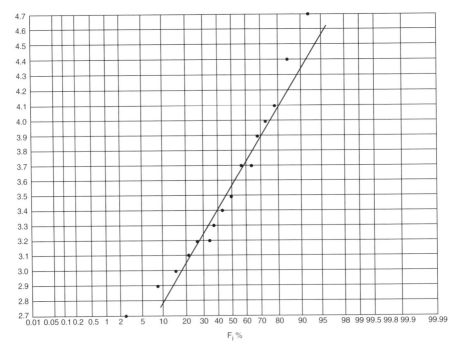

Figure 3.15. Mean of ten casts of dice.

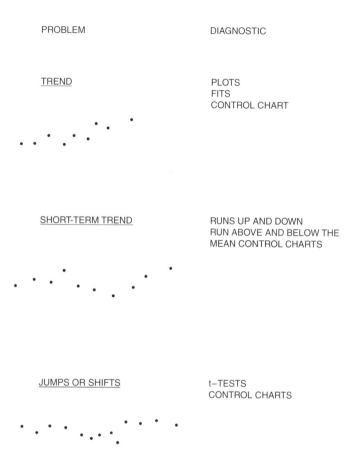

Figure 3.16. Gross deviations from randomness.

mentioned that MINITAB statistical software would be discussed due to its wide range of features for metrologists with basic to intermediate statistical knowledge. At the time of this books printing, the current version of MINITAB was Release 14. The MINITAB website, www.minitab.com, offers a free 30 day trial version of the software. The user is able to install and use a full-scale version of MINITAB for 30 days before making the commitment to purchase the software. A number of user manuals are available which detail MINITAB's use [5].

MINITAB software appears as a divided screen when opened. The upper half of the screen is a session window, with pull-down menu headings. The lower half of the screen is a spreadsheet which displays columns for any data being manipulated. Data manipulation is possible by either typing the commands for specific functions using a command editor window or by using the pull-down menus. Users new to MINITAB (and thus not familiar with MINITAB commands) will find the pull-down menus much easier to manipulate. The remainder of the discussion in this book focuses on data manipulation using the pull-down menus. Familiarity with pull-down software menus is assumed.

MINITAB Example

One useful feature of MINITAB software is that it has several datasets included as part of the installation. The user is able to select a dataset and perform various statistical analyses on it. One such dataset details the amount of radiation detected in an experimental chamber using 4 different measuring devices: filters, membranes, open cups and badges. The dataset is called 'radon.mtw', and is shown in Table 3.3. The menu commands to access the dataset are as follows:

Under File, Open Worksheet
• double click on 'radon.MTW', click OK

This will open the radiation dataset, so that the user sees the data in the spreadsheet portion of the screen.

Table 3.3. Radiation Dataset from MINITAB

Filter	Membrane	Open Cup	Badge
26	45	36	21
21	33	34	23
16	26	33	27
28	46	29	29
27	25	30	24
19	32	36	26
21	33	28	27
26	22	34	26
25	30	34	24
16	26	33	25
20	37	33	31
25	34	29	28
25	44	28	30
21	43	27	30
23	35	27	31
26	38	30	29
21	38	29	29
23	45	28	29
17	39	31	30
25	39	27	28

Taken from MINITAB

The user may compute descriptive statistics such as the mean and the median for the filter measurement (or any of the other measurement types) by selecting the following menu commands:

Under Stat, Basic Statistics, Display Descriptive Statistics
• double click on C1 Filter to add it to the **Variables** field, click OK

The mean and the median (shown in bold) are printed for the filter data in the session window.

MINITAB Output

Descriptive Statistics: Filter

Variable	N	Mean	Median	TrMean	StDev	SE Mean
Filter	20	**22.550**	**23.000**	22.611	3.663	0.819

Variable	Minimum	Maximum	Q1	Q3
Filter	16.000	28.000	20.250	25.750

Taken from MINITAB

Only certain descriptive statistics are available for display. To compute the mode, skewness, and kurtosis, one uses the same commands as above, but rather than displaying, storing is selected:

Under Stat, Basic Statistics, Store Descriptive Statistics
• double click on C1 Filter to add it to the **Variables** field
• under **Statistics**, select Mean, Median, Skewness, Kurtosis, click OK, click OK

The mean, median, skewness, and kurtosis are now stored as columns C5-C8 in the spreadsheet. The number of observations that the calculations are based on (in this case 20) is also provided:

MINITAB Output

Mean1	Median1	Skewness1	Kurtosis1	N1
22.55	23	-0.448526	-0.851983	20

Taken from MINITAB

Storing the columns, rather than just displaying them, enables the results to be used in future calculations. To see if the radiation measurements collected using the membrane have a normal distribution, the following menu commands are chosen:

Under Graph, Probability Plot
• click **Single**, click OK
• double click on C2 Membrane to add it to the **Graph Variables** field
• under **Labels**, add a title in the **Title** field, click OK, click OK

The resulting normal probability plot of the data is shown in Figure 3.17:

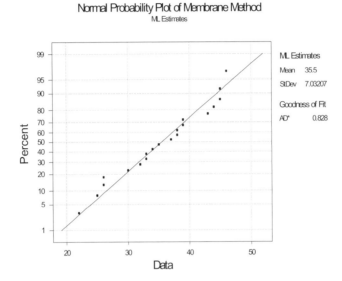

Figure 3.17. Normal probability plot-membrane method.

The data points form an approximately straight line, indicating no gross deviations from normality. MINITAB also offers features for adding 95% confidence intervals to the graph (see the Options button for more information). The graph above uses the normal distribution for the plot, which is the default. Options exist to select other distributions, including log normal.

EXERCISES

3-1. Describe the basic objectives of statistical analysis of data.

3-2. Define: precision, bias, and blunder. Explain how each contributes to the overall accuracy of a measured value.

3-3. What is meant by a normal distribution? Explain using a sketch, and show five variations of such a distribution.

3-4. Given the following data set:

20.60, 20.65, 20.35, 20.45, 20.70, 20.20, 20.50, 20.37, 20.65, 20.50

    a.   Plot on arithmetic probability paper to test for normal distribution. Plot using the computer software of your choice.

    b.   Given the mean, $\overline{X} = 20.497$ and $\sigma = 0.159$, calculate the skewness factor, g_1, and the kurtosis factor, g_2.

    c.   Repeat calculations using statistical software.

3-5. Given the following data set:

8.4, 7.4, 9.6, 10.6, 9.0, 6.9, 8.0, 13.0, 5.9, 16.5

Use a computer generated probability plot to decide on the nature of the distribution, i.e., whether it appears to be closer to normal or to log normal.

3-6. Describe the basic requirements that a data set must meet before it may be subjected to statistical analysis. How can one be assured that these requirements are met in a particular situation? Discuss gross violations of these requirements. If possible, give examples based on your own experience or your observations on data obtained by others.

3-7. Explain the difference between the terms uncertainty and error as used in data analysis. How can erroneous data affect statistical conclusions? Describe the advice that you might give to a colleague or client on causes of error in data and how to screen for such. Use a practical example, as possible.

REFERENCES

[1]   Eisenhart, C. "Realistic Evaluation of the Precision and Accuracy of Instrument Calibration Systems," Reprinted in *Precision Measurement and Calibration: Statistical Concepts and Procedures*, Ku, H.H. Ed. NBS Spec. Publ. 300, Vol. 1, (Gaithersburg, MD: National Institute of Standards and Technology).

[2]   TEAM, P.O. Box 25, Tamworth, NH 03886, Tel. (603) 323-8843. Probability paper also may be available locally in drafting supply stores.

[3]   Ku, H.H., "A Users' Guide to OMNITAB Command "STATISTICAL ANALYSIS," NBS Technical Note 756, (Gaithersburg, MD: National Institute of Standards and Technology, 1973).

[4]   D'Agostino, R.B., and G.L. Tietjen. *Bimetrika* 60: 172 (1973).

[5]   *MINITAB User's Guide 2: Data Analysis and Quality Tools,* Release 14 for Windows (Minitab Inc. 2003, p 9-7)

CHAPTER **4**

Statistical Calculations

This chapter reviews the procedures most commonly used to estimate the statistical parameters of measurement data upon which statistically based decisions about scientific and engineering data are made. A thorough understanding of these procedures will not only permit one to use them intelligently for data analysis but should also prove to be useful for better planning and execution of measurement programs.

INTRODUCTION

Chapter 3 contained a discussion on characterization of a data set by the mean and the standard deviation. In the following discussion it will be assumed that there is a data set that is worthy of statistical description, meeting the requirements of Chapter 3. Most often it will be what may be a small portion of a much larger set that could have been obtained if time, money, and other considerations had permitted. This smaller set will be called a *sample* of a *population* (see Chapter 1). Ordinarily, measurements are made not to characterize a sample but rather to characterize the population from which it may have been taken.

The larger body (when it contains all of the possible units) is called the population because much of statistical theory arose out of studies of people. Population means all of the entire universe that is under study. The population of the United States means all of the people that live in this country. If one wanted to obtain information about some characteristic of the population, it would be best to examine every individual.

```
POPULATION
Means All
SAMPLE
Means Some
```

47

This usually is not feasible, so less than all, i.e., a sample, will be studied. If this is a random sample, there is hope that the results of the study will give unbiased information about the population. Obviously, a biased sample, e.g., examination of people from only a small region, would result in a biased conclusion about the question investigated. The same is true for any population and any sample.

Chemists may think of the word sample too narrowly as being a portion of some material that they may analyze, although this can be a proper way to use the term. However, the word may be used generically to mean a portion of something larger that is used to provide some needed information about the population of which the sample is a member.

In measurement, there is a similar situation. One could sit down and make an infinite number of measurements on a stable specimen of material and use the results to characterize the precision of the measurement process. This is impossible, so a much smaller number of measurements must be used for this purpose, that is to say a sample must be used. Again, this sample will have to be an unbiased sample of the performance of the measurement process (i.e., typical performance) or else a biased opinion will be obtained of the ability of the process to perform.

THE MEAN, VARIANCE, AND STANDARD DEVIATION

How could one calculate the statistical parameters of a simple data set? The formulas to be used in the case of a normal distribution are summarized in Figure 4.1. If all of the possible measurements that could have been made were available, one could evaluate the average or mean value, μ, of the population, also designated as m by some statisticians. The dispersion of the individual values around the mean is measured by the standard deviation, σ, as discussed earlier. Another measure of variability is the variance, which is equal to σ^2. The symbol V is sometimes used to designate variance.

Ordinarily one is not dealing with a population, but rather with a sample of n individuals of the population. The individual measured values may be indicated by the symbols X_1, X_2, X_n.

The sample mean, $\overline{X}$ (called X bar), calculated as shown in the figure, is hopefully a good estimate of the population mean–that is why the measurements were made in the first place! One can calculate the sample standard deviation, s, using the formula shown in the figure. Likewise one can calculate the sample variance, s^2, as shown. Of course, one can use a simple calculator to do this, as indicated by PUSH BUTTON. This convenience is good because arithmetic is hard work and one may make mistakes. However, one should make a few calculations by the formula just to understand and appreciate what the calculator is doing.

Remember that s is an estimate of the standard deviation of the population and that it is not σ. It is often called "the standard deviation", maybe because the term 'estimate of the standard deviation' is cumbersome. It is, of course, the sample-based

Population	Sample
All measurements	Sample of n measurements
	$x_1, x_2, \ldots x_n$
Mean μ, m	$\bar{x} = \Sigma x_i/n$
Standard Deviation, σ	$s = \sqrt{\dfrac{\Sigma(x_i - \bar{x})^2}{n-1}}$
Variance σ^2 (V)	$s2 = \dfrac{\Sigma(x_i - \bar{x})^2}{n-1}$
	or
	PUSH BUTTON

df = v = degrees of freedom, often n − 1

$s_{\bar{x}}$ = s/$\sqrt{n}$ = standard deviation or standard error of mean

s/$\bar{x}$ = coefficient of variation, cv

s/$\bar{x}$ × 100 = relative standard deviation, RSD

Figure 4.1. Population values and sample estimates.

standard deviation but that term is also cumbersome. The standard deviation and its estimates always have the same units as those for X. When considering variability, a dimensionless quantity, the coefficient of variation, cv, is frequently encountered. It is simply

$$cv = s/\overline{X}$$

If one knows cv and the level, X, s can be calculated.

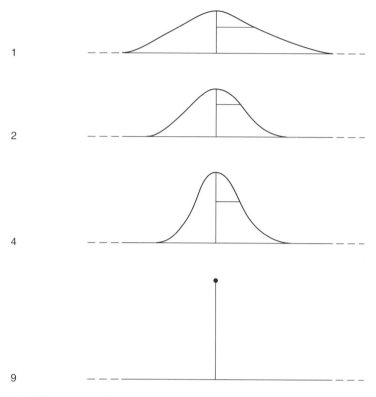

Figure 4.2. Distribution of means.

Another term frequently used is called the relative standard deviation, RSD, and it is calculated as

$$RSD = cv \times 100$$

The relative standard deviation is thus expressed as a percent. There could be room for confusion when results are reported on a percentage basis, as the percentage of sulfur in a coal sample, for example. Here the value for s could be in units of percent. In such cases, one can make the distinction by using the terms % relative and % absolute.

One other piece of information is appropriate at this point. The value for s calculated above is for that of individuals about the mean. Knowing this s, how would one expect the means of sets of n measurements to vary? This quantity is, of course, the standard deviation of the mean (also called the standard error of the mean) and is given by

$$s_{\bar{x}} = s_x / \sqrt{n} = \text{standard error}$$

This is an important relationship to remember. It tells one that the precision of the mean increases inversely proportionally to the square root of the number of measurements. Thus, the standard deviation (standard error) for the mean of 16 measurements for which $s_X = 4$ would be the same as the standard deviation for a single measurement for which $s_X = 1$. To review, s_X is the standard deviation for individual values; $s_{\bar{X}}$ is the standard deviation for means of n values.

Figure 4.2 illustrates distributions of the standard deviation of the mean. In the top of the figure, the dispersion of individuals having a standard deviation of 1 is shown. Just below it, the dispersion of the means of two such measurements is shown. They have a standard deviation of $1\sqrt{2} = 0.707$. Below this is the dispersion of means of four measurements, having a standard deviation of $1\sqrt{4} = 0.500$. It will be noted that all of the curves must have the same area, namely 1, hence the height is inversely related to the narrowness, i.e., to the precision of the process. With the above information, the reader may wish to sketch the distribution of means of nine measurements in the space provided at the bottom of the figure.

Example The following calculations are presented to illustrate the above points.

Estimation of Standard Deviation from a Data Set

X_i	$(X_i - \bar{X})$	$(X_i - \bar{X})^2$
11	0	0
15	4	16
7	−4	16
12	1	1
14	3	9
9	−2	4
10	−1	1
8	−3	9
13	2	4
11	0	0
Σ 110		60

$\bar{X} = 110/10 = 11.0$ $s = \sqrt{(60/9)} = 2.58$ $s^2 = 6.66$

$cv = 2.58/11 = 0.235$

$RSD = 0.235 \times 100 = 23.5\%$

$s_{\bar{X}} = 2.58/\sqrt{10} = 0.815$

$s_{\bar{X}}^2 = 6.66/10 = 0.666$, or $(0.815)^2 = 0.666$

MINITAB offers computation of the standard error of the mean in addition to computation of the mean and standard deviation by selecting the following menu commands:

Under Stat, Basic Statistics, Store Descriptive Statistics
* under **Statistics**, select mean, SE of mean, standard deviation, variance, click OK, click OK

DEGREES OF FREEDOM

A very important concept in statistics is called the "degrees of freedom", yet it is difficult to visualize. Actually, the term is used several ways. As used in the techniques described in this book, it may be considered to refer to the number of independent estimates of the statistic of interest that could be obtained from a given data set.

For example, if one had ten measurements, they could be used in pairs to obtain nine different estimates of the mean. When one tried to make a tenth estimate, it would be found that one of the pairs already used was involved. Of course, all of the means could be put together to obtain a mean for the entire data set. One would say that the mean had been determined with 9 degrees of freedom. The standard deviation could be estimated using pairs of data in a similar manner. The standard deviation of the above data set would have been estimated with 9 degrees of freedom. If only two data points were available, only one mean and one standard deviation could be estimated and the degrees of freedom would be $df = 2 - 1 = 1$.

In general, the degrees of freedom, also designated by the lower case Greek letter nu, are calculated by

$$df = n - number\ of\ parameters\ estimated$$

For means and standard deviations, $df = n - 1$. If a straight line is fitted to a set of data, the equation has two constants and they would be determined with $n - 2$ degrees of freedom, where n is the number of data points used to evaluate the constants. In the above, it is assumed that all of the data points are mutually independent of one another! If this is not true, then the estimates are based upon a lesser number of degrees of freedom. It will be seen in Chapter 5 that the degrees of freedom is one of the parameters used to define the values for t and F that are used to make statistical estimates such as confidence limits and decisions about the significance of apparent measurement differences.

USING DUPLICATE MEASUREMENTS TO ESTIMATE A STANDARD DEVIATION

Sometimes it is of advantage to estimate standard deviations from sets of duplicate measurements of a single sample or of each of several samples of similar composition and hence having essentially the same precision of measurement. The procedure for use is called the pooled standard deviation estimate, s_p, since an estimate of the standard deviation could have been made from each of the sets of duplicates, and this information is pooled in a single calculation.

$$s_p = \sqrt{(\Sigma d_i^2/2k)}$$

The relationship to use is where d_i represents the difference of a duplicate measurement and k is the number of sets of duplicates that were measured. s_p will be estimated with k degrees of freedom. Notice that the above expression is a special case of that given in the section Pooled Statistical Estimates of this chapter.

Example

Same Sample Measured on Various Occasions

Experiment No.	X_1	X_2	d	d^2
1	14.5	14.2	.3	.09
2	14.8	14.9	.1	.01
3	14.2	14.8	.6	.36
4	14.7	14.1	.6	.36
5	14.9	15.1	.2	.04
6	14.4	14.7	.3	.09
7	14.3	14.5	.2	.04
Σ				0.99

$$k = 7 \qquad s = \sqrt{(0.99/14)} = 0.27 \qquad df = 7$$

Example

Different Samples

Sample No.	X_1	X_2	d	d^2
1	14.7	14.4	.3	.09
2	20.1	20.5	.4	.16
3	16.7	16.5	.2	.04
4	19.3	19.9	.6	.36
5	15.0	14.7	.3	.09
				0.74

$$k = 5 \qquad s = \sqrt{(0.74/10)} = 0.27 \qquad df = 5$$

It will be noticed from the above examples that one does not have to use the same sample for estimation of the standard deviation. In fact, the level of the samples used can vary, as long as it is believed that the standard deviation should be essentially the same for the measurement of all of the samples included in the group.

USING THE RANGE TO ESTIMATE THE STANDARD DEVIATION

There is a third way that is commonly used to estimate the standard deviation, and that is from the range of a set of measurements. The range is the difference between the largest and the smallest value in a set. In a set of duplicates, it is simply the absolute difference of the two values. The range (divided by the appropriate factor) is a good estimate of the standard deviation for small data sets (say up to 6) but the classical calculation is better for larger sets.

The idea is that the range is inversely related to the standard deviation by the constant d_2^* which can be found in statistical tables. Table A.1 in Appendix A is such a table. A more extensive one will be found in the reference below the table. To use it, one calculates the average range, $\overline{R}$, from several sets of data, i.e.,

$$\overline{R} = (R_1 + R_2 + \ldots + R_k)/k$$
$$s = \overline{R}/d_2^*$$

In the table, the size of sample means the number in each set, i.e., 2 for duplicates, 3 for triplicates, etc. The number of sets of samples is k. Thus, for 10 sets of triplicates, d_2^* would be 1.72.

The number of degrees of freedom associated with the estimate is also shown. In the example above, df = 18.4. That does not sound like n – any number, does it? That is because of hidden replication in the sets. Do not worry about that. Leave that worry to the professional statistician who will gladly explain it to you! The value for df to be used in statistical calculations is rounded down to the next smaller integer.

Estimation of Standard Deviation from Average Range
Triplicate Measurements

Data Set			R
14.5,	14.2,	14.4	.3
14.8,	14.9,	14.9	.1
14.8,	14.2,	14.4	.6
14.7,	14.3,	14.1	.6
15.1,	14.9,	15.0	.2
14.7,	14.4,	14.6	.3
14.3,	14.6,	14.5	.3

$$\overline{R} = \frac{.3 + .1 + .6 + .6 + .2 + .3 + .3}{7} = 0.343$$

$$d_2^* = 1.73 \qquad s = 0.343/1.73 = 0.198$$

$$df = 12 \text{ (by interpolation)}$$

POOLED STATISTICAL ESTIMATES

Because of the variability of data, one would expect to get somewhat different values for estimates of means and standard deviations on each data set of the population that is used. How can one combine such information to obtain a better estimate, based on a larger data base? In the case of means, the answer is obvious – one simply calculates a grand average, i.e., an average of the averages. This is correct if all of the sets so combined have peer status, that is, each is equally precise and composed of the same number of data points. In Chapter 6, a discussion is presented on how this approach should be modified when these assumptions are not valid. What about estimates of standard deviations? First they must all be estimates of the same standard deviation, that is, one should not combine standard deviation estimates for several different methods of measurement, for example. It will be shown later how one can use what is called an F test (Chapter 5), and also the Cochran test (Chapter 6), to decide on the significance of apparent differences among s values. However, for the present purpose, it will be assumed that one has k different estimates of the same standard deviation that one wishes to pool.

One should not average s values but rather one can average variances to obtain an average of the variances. Because the various estimates may not have been based on the same number of degrees of freedom, they are weighted according to their respective degrees of freedom, df_i, so that those based on larger numbers of degrees of freedom get greater weight. The relationship to be used is

$$s_p = \sqrt{\left(\frac{s_1^2 \times df_1 + s_2^2 \times df_2 + \ldots + s_k^2 \times df_k}{df_1 + df_2 + \ldots + df_k} \right)}$$

The value for s_p will be based on Σ df degrees of freedom. Given the following sets of data:

Set No.	s	n	df	s^2	$df \times s^2$
1	1.35	10	9	1.822	16.40
2	2.00	7	6	4.000	24.00
3	2.45	6	5	6.002	30.01
4	1.55	12	11	2.402	26.43

$$s_p = \sqrt{\left(\frac{16.40 + 24.00 + 30.01 + 26.43}{9 + 6 + 5 + 11}\right)} = 1.77$$

$$df = 31$$

SIMPLE ANALYSIS OF VARIANCE

In what has been discussed so far, it has been assumed that all of the observed variance for a set of data comes from one source. This could be due to the variability of measuring a homogeneous sample of material, or it could be due to the variability of a heterogeneous material, when using a method of measurement that was essentially invariant, relatively speaking. However, what is the case when several sources of variance are present? Such a situation often happens in the real world when both the material analyzed and the method of measurement have significant variances, for example.

In general, when sources of variability are linearly related (independent and uncorrelated), the respective variances are additive and the total variance, s_T^2, is expressed as

$$s_T^2 = s_1^2 + s_2^2 + \cdots + s_k^2$$

Sometimes, it is possible to devise experiments to evaluate the magnitudes of the several sources of variance. The following discusses a very simple but frequently occurring situation, namely that in which variances due both to a material and to its measurement are of concern.

Let s_m^2 = the variance of the material

s_a^2 = the variance of the method of measurement

If replicate measurements are made on each of a number of suitably selected samples, the results of the replicate measurements on samples of a material may be used to estimate the precision of measurement (and also the variance) while the variability of the averages for the several samples is due to both sources of variance. Knowing the measurement variance, the material variance can be calculated.

This may be illustrated by considering the duplicate measurement of k samples of material. This very practical procedure involves a minimum amount of work when both variances need to be estimated. The measurements to be made and the calculations to be done are as follows:

1.   Calculate $\overline{R}$
2.   Calculate

$$s_a = \overline{R} / d_2^*$$

or alternatively

$$s_a = \sqrt{(\Sigma d^2 / 2k)}$$

s_a is estimated with $df = k$

3.   Calculate the standard deviation of the means in the last column by conventional method for a set of values. Call this s_T

4. $$s_T^2 = s_m^2 + s_a^2 / 2$$

5.   Substitute value for s_a from step 2
6.   Calculate s_m (s_m is estimated with $df = k - 1$)

Sample No.	Measured Values		R or d	Mean
1	X_1'	X_1''	R_1 or d_1	X_1
2	X_2'	X_2''	R_2 or d_2	X_2
.	.	.	. .	.
.	.	.	. .	.
.	.	.	. .	.
k	X_k'	X_k''	R_k or d_k	$\overline{X}_k$

The value of s_a^2 is divided by 2 because each value for $\overline{X}$ is the mean of two measurements. The value for s_a was ob-tained using the range. The difference-of-duplicates formula could be used, as well. Of course, if the value for s_a is already known from another source of information, one simply makes single measurements on k samples, as was illustrated in Figure 3.4A, and calculates s, which is really s_T. Then s_m is calculated by the expression

$$s_m = \sqrt{(s_T^2 - s_a^2)}$$

Example

Sample No.	Measured Values		d	$\overline{X}$
1	16.0	16.2	.2	16.10
2	14.9	15.3	.4	15.10
3	15.8	15.8	0	15.80
4	16.1	15.9	.2	16.00
5	15.7	15.2	.5	15.45
6	15.9	15.7	.2	15.80
7	15.1	15.4	.3	15.25

$$\overline{R} = (.2 + .4 + 0 + .2 + .5 + .2 + .3)/7 + 0.257$$

$$d_2^* = 1.18 \quad df = 6, \quad s_{a_1} = 0.218$$

$$s_{\overline{x}} = s_T = 0.381, s_m = \sqrt{[(.381)^2 - (.218)^2/2]} = 0.348$$

The procedure described above may be depicted by the following scheme:

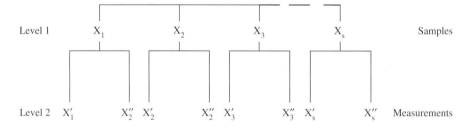

| Level 1 | X_1 | | X_2 | | X_3 | | X_s | Samples |

| Level 2 | X_1' | | X_2'' X_2' | | X_2'' X_3' | | X_3'' X_s' | X_s'' Measurements |

This is called a nested design because the measurement variance, in level 2, may be considered to be nested in the total variance of level 1. This nested design makes use of duplicates as a means to evaluate the variance of levels below level 1. However, any number of replicates could be used, e.g., triplicate measurements, to give estimates of variance (and hence standard deviations) based on a larger number of degrees of freedom. The use of duplicates is often a reasonable choice, based on practical considerations.

Multilevel nested designs may be used to evaluate variance due to several variables. A three-level nested design will be described that permits the estimation of variances due to between sample, within sample, and measurement variability, respectively. Other sets of variables such as between laboratory, within laboratory, and between analysts could be investigated using the same design. The extension of the design to evaluate additional levels should be obvious with some contemplation by the reader.

The nested design based on duplicates, is as follows:

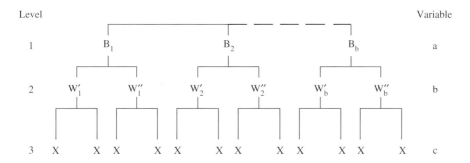

In the design shown above, B represents samples taken randomly from a production lot or from a sampling location. The total number taken is b. W represents duplicate samples randomly taken from within each sample, B. The total number taken is 2b. X represents a measured value of each one of the samples, W. The total number of measurements is 4b. The values, X, are the raw data from which the values for W and B may be calculated. Note that in design terminology, the variable which is the actual measurement is referred to as the response, whereas the classifying variables are referred to as factors.

The computational procedure that may be used is as follows:

1. To calculate $s_3{}^2$ the variance at level 3 (in this case, it is the variance of the measurement process).

   a. Calculate the differences, d_3, of the pairs of values at Level 3

   b. $s_3^2 = s_m^2 = \Sigma d_3^2/4b$ df = 2b (because there are 2b sets of differences).

2. To estimate $s_w{}^2$ the variance of samples within the bags

   a. Calculate the values for level 2 from the data of level 3, e.g., the first value at level 2 is the average of the first two values at level 3, etc.

   b. Calculate the differences, d_2, of the pairs of values at Level 2.

   c. Estimate s_2^2, the variance at level 2

$$s_2^2 = \Sigma d_2^2/2b$$

d. $s_W^2 = s_2^2 - s_m^2/2$ df = b (because there are b sets of differences). Note that s_m^2 is divided by 2 because each value of W is based on 2 measured values.

3.   To estimate $s_B{}^2$, the variance of the samples in the lot.

    a.   Calculate the values for B, e.g., the first value for level 1 is the average of the first two values of level 2, etc.

    b.   Estimate s_1^2, the variance of level 1, as follows:

$$s_1^2 = \Sigma(B - \overline{B})^2/b - 1$$

    c.   Estimate $s_B^2 = s_1^2 - s_2^2/2$

The value for s_2' is divided by two because two values for W are used to calculate each value for B. Alternatively,

$$s_B^2 = s_1^2 - s_w^2/2 - s_m^2/4 \quad df = b - 1$$

Example   Seven bags of material were randomly selected from a manufactured lot. Two subsamples were randomly selected from each bag. Duplicate measurements were made on each of these subsamples. Estimate the variance of measurement, the variance of the composition of the material within bags, and the variance of material between bags. The data have been coded (see the section on Computations) by subtraction of 10.00 from each measured value and are tabulated using the format of Table 4.1. The coded data are listed in column 1.

Calculations*

1. $s_3^2 = s_m^2 = .0624/28 = 0.002229$ $df = 14$
     $s_m = .0472$ $df = 14$

2. $s_2^2 = .1452/14 = .010371$
     $s_w^2 = .010371 - .002229/2 = 0.0092565$ $df = 7$
     $s_w = .09621$ $df = 7$

Data, Level 3, X*	d₃	Level 2	d₂	Level 1
.78				
	.04	.80		
.82				
			.20	.70
.56				
	.08	.60		
.64				
. . . .				
.50				
	.10	.45		
.40				
			.10	.40
.36				
	.02	.35		
.34				
. . . .				
.56				
	.08	.52		
.48				
			.16	.60
.62				
	.12	.68		
.74				
. . . .				
.78				
	.06	.75		
.72				
			.10	.80
.84				
	.02	.85		
. . . .				
.86				
.84				
	.10	.79		
.74				
			.14	.72
.62				
	.06	.65		
.68				
. . . .				
.47				
	.00	.47		
.47				
			.16	.55
.66				
	.06	.63		
.60				
. . . .				
.37				
	.04	.39		
.41				
			.12	.45
.50				
	.02	.51		
.52				

Table 4.1. Format for Tabulation of Data Used in Estimation of Variance at Three Levels, Using a Nested Design Involving Duplicates

Original Data Level 3	d_3	Level 2	d_2	Level 1
————————				
	—————	—————		
————————			—————	—————
	—————	—————		
————————				
· · · · · · · · · · · ·				
————————				
	—————	—————		
————————				
			—————	—————
————————				
	—————	—————		
————————				
· · · · · · · · · · · ·				
————————				
	—————	—————		
————————				
			—————	—————
————————				
	—————	—————		
————————				
· · · · · · · · · · · ·				
————————				
	—————	—————		
————————				
			—————	—————
————————				
	—————	—————		
————————				
· · · · · · · · · · · ·				
————————				
	—————	—————		
————————				
			—————	—————
————————				
	—————	—————		
————————				
· · · · · · · · · · · ·				
————————				
	—————	—————		
————————			—————	—————
	—————	—————		
————————				

The format is based on the use of a sample size of seven at level 1. It may be expanded or contracted to accommodate larger or smaller samples.

3. $s_1^2 = .021557 \ \overline{B} = .603 \ (coded) \ \overline{B} = 10.603$

$s_B^2 = s_1^2 - s_2^2 / 2$

Note that the second term is divided by 2 because 2 values of W are required to calculate each value for B. Alternatively,

$s_B^2 = s_1^2 - s_w^2/2 - s_m^2/4$

$s_B^2 = .021557 - .0092565/2 - .002229/4$

$s_B^2 = 0.0163715 \quad df = 6$

$s_B = 0.1280 \quad df = 6$

Note: The coded and actual variances and standard deviations are identical. The mean is obtained by adding 10.00 to the coded mean. For this dataset, the level 3 variable is the response and the level 2 and level 1 variables are the factors.

In MINITAB, one can select options for nested analysis of variance to perform these calculations. The data must first be entered in the form shown in Table 4.2 to indicate samples and subsamples.

Table 4.2.   Material Bag Dataset from MINITAB

Measurement	Sample	Subsample
.78	1	1
.82	1	1
.56	1	2
.64	1	2
⋮	⋮	⋮
.60	6	2
.37	7	1
.41	7	1
.50	7	2
.52	7	2

Using the following MINITAB menu commands:

Under Stat, ANOVA, Fully Nested ANOVA
- double click on C1 Measurement to add it to the **Responses** field
- double click on C2 Sample and C3 Subsample to add them to the **Factors** field, click OK

the output below is produced.

MINITAB Output

```
Nested ANOVA: Measurement versus Sample, Subsample
Analysis of Variance for Measurem
Source      DF           SS          MS      F        P
Sample      6         0.5174      0.0862   4.157   0.042
Subsampl    7         0.1452      0.0207   9.308   0.000
Error      14         0.0312      0.0022
Total      27         0.6938

Variance Components
Source      Var Comp.   % of Total      StDev
Sample        0.016        58.77        0.128
Subsampl      0.009        33.23        0.096
Error         0.002         8.00        0.047
Total         0.028                     0.167

Expected Mean Squares
 1 Sample     1.00(3) +   2.00(2) +   4.00(1)
 2 Subsampl   1.00(3) +   2.00(2)
 3 Error      1.00(3)
```

Taken from MINITAB

The measurement standard deviation is shown under error, the between bag standard deviation under sample, and the within bag standard deviation under subsample.

LOG NORMAL STATISTICS

The statistical calculations given above apply **only** to normal distributions. However, data may sometimes be transformed to achieve normality. The logarithmic transformation is a very common example of such. What should be done in the case of log normal distributions? The answer is simple. One merely converts all of the data, logarithmically, and handles the logs of the data as if they were a set of normal data. After all calculations are made, the antilogs are taken to obtain the mean. The log standard deviation has meaning but its antilog does not. One needs to make calculations such as confidence limits first with the logs and then convert the results to the form of the original data. This will be discussed further in Chapter 5.

The following example is presented to illustrate the estimation of the mean and the standard deviation of a set of log-normally distributed data.

Example   Consider the log normal data set given in Chapter 3:

X	log X
7.0	.84510
9.9	.99564

Avg of $\log \overline{X} = 1.25081$

$\overline{X} = 17.8$

s of $\log \overline{X} = .32412$

X	log X
12.5	1.09691
24.0	1.38021
70.0	1.84510
7.0	.84510
16.0	1.20412
30.5	1.40430
19.0	1.27875
41.0	1.61278

(Note that $\overline{X}$ is exactly the geometric mean)

The limits of the population corresponding to $\pm\sigma$, encompassing two thirds of the data, are as follows:

$$1.25081 - .32412 = 0.92669, \text{antilog} = 8.4$$
$$1.25081 - .32412 = 1.57493, \text{antilog} = 37.6$$

The procedures for use in making statistical decisions about such data will be discussed in Chapter 5.

MINIMUM REPORTING STATISTICS

Assuming that one wishes to report the results of measurement, what is the minimum necessary amount of statistical information that can be meaningful? The answer, in the case of normally distributed data, is the values for the following three parameters:

$\overline{X}$	The mean of the dat set
s	The estimate of the standard deviation
n	The number of independent data points

Instead of n, one can report df, the number of degrees of freedom.

All of the above must be reported, else the ones reported are essentially meaningless. How good an estimate $\overline{X}$ is of μ depends on the number of data points. Obviously, more is always better! The value of s is also needed because, as it will be seen, the confidence interval for $\overline{X}$ is directly proportional to the value for s. Also, how much confidence one has in s will depend on the value of n, or more properly on df. Yet one often sees data reported in the literature that is incomplete in the above respects.

Given the entire data set or the information described above, one can make many different kinds of decisions about the results of measurements. Each will depend on the kind of decisions that need to be made, based on the data. The most important ones will be considered in the following Chapters. When these relatively simple techniques have been mastered, one will be in a position to statistically analyze many kinds of measurement data.

COMPUTATIONS

Data are sometimes coded before computations are made to facilitate the arithmetic involved. The two possible coding operations used are

a.   Multiplication (or its inverse, division) to change the order of magnitude of the original numbers.
b.   Addition (or its inverse subtraction) to reduce the number of figures that are carried along in the computations.

After computations are completed, the data may need to be decoded, which is to say it may need to be changed back into its original units.

Natrella [1] lists the following five rules that should be observed for coding and decoding:

1.   The whole set of data must be treated alike.
2.   Either operation (a) or (b), or both together may be used to make the data more tractable.
3.   Keep careful notes on the coding process.
4.   Perform the desired computation on the coded data.
5.   Decode the computed results as necessary.

The mean is affected by any coding operation. Therefore, the inverse operation must be applied to decode the results. A standard deviation estimate is affected by multiplication or division but not by addition or substraction. Hence these results need to be decoded accordingly. A variance computed on coded data must be multiplied by the square of the coding factor if division is used in coding: or divided by the square of the coding factor if multiplication is used in coding. As long as no rounding is involved in the coding process, there is no loss of significant figures due to coding.

Rounding of data is a question that should be carefully considered. Measurement data are inherently rounded by the measurement process and may be rounded further by the observer [2]. This is often because of the great emphasis on not to report more figures than are reliable, which may lead observers to record less than enough figures. Rounding should be the last operation in a computation. It may be disastrous to round data before using it for computational purposes.

The following computational rules are given to refresh the memory of the reader on this matter.

1. *Addition*. When several numbers are added, the sum should be rounded to the number of decimal places no greater than in the addend, which has the smallest number of decimal places. For example:

$$
\begin{array}{r}
4.10 \\
.0001 \\
.50 \\
\hline
4.6001
\end{array}
\quad \text{rounded to } 4.60
$$

2. *Subtraction*. Follow the same rule as addition, rounding the final result. For example:

$$
\begin{array}{r}
10.30 \\
-.126 \\
\hline
10.174
\end{array}
\quad \text{rounded to } 10.17
$$

3. *Multiplication*. Round off the result to the number of significant figures contained in the smallest multiplicand. For example:

$$10 \times 25.3 \times 1.257 = 318.021 \text{ rounded to } 3.2 \times 10^2$$

4. *Division*. Round off the result to the number of significant figures in the dividend or divisor, whichever is smaller. For example:

$$5412 \div 1.23456 = 4383.748 \text{ rounded to } 4384$$

5. *Powers and Roots*. Round off to the number of figures in the original data. For example:

$$\sqrt{12.3} = 3.5071355 \text{ rounded to } 3.51$$
$$(12.3)^2 = 151.29 \text{ rounded to } 151$$

6. *Logarithms*. The antilog is correct to $n - 1$ significant figures, where n is the number of figures in the mantissa.

In rounding, when the digit immediately following the figure to be retained is greater than 5, the retained figure is increased by one and conversely. When the succeeding figure is exactly 5, and the retained figure is even, it is left unchanged. Otherwise, add one to make the retained figure even. Exponential notation should be used for numbers of large magnitude containing a small number of significant figures. Thus one should write 2.5×10^3 to express 2500 to two significant figures.

For complex operations, the final result is rounded to the number of significant figures of the operator containing the least number. Remember that rounding is the last operation and it is proper and commendable to keep more figures than are significant in the intermediate operations. The author likes to keep at least two additional figures if they are available. This subject is discussed further in Chapter 5.

ONE LAST THING TO REMEMBER

Remember that $\overline{X}$ and s are estimates of the population statistics μ and σ, respectively. They pertain only to a specific population. One must specify or otherwise make clear what the population is for which they are the statistical parameters. Otherwise, they are merely numbers and even meaningless numbers. If anything changes about the population, one would expect the statistics to change also and hence they should be re-evaluated. The above statements are especially pertinent to measurement. If a measurement system is changed in any way, such as installation of a different detector or using a different solvent, one will probably need to re-estimate the standard deviation, at the least. In fact, such changes are often made to 'improve the precision', in which case one would be disappointed if the standard deviation had not decreased. Likewise, the precision of the results produced by two analysts using the same equipment and methodology could be expected to differ, if operator variance is of any significance.

EXERCISES

4-1. Discuss the difference between sample and population statistics.

4-2. What is meant by an unbiased sample? Why is this important?

4-3. Given the following data set:

51, 55, 47, 52, 54, 49, 50, 48, 53, 51

a. Compute the sample mean, standard deviation, and degrees of freedom for s.
b. Compute the coefficient of variation.
c. Compute the relative standard deviation.
d. Use MINITAB software for these computations.

4-4. Given the data set:

151, 155, 147, 152, 154, 149, 150, 148, 153, 151

a. Compute the mean, s, cv, and RSD.
b. What do you conclude by comparing the statistics of Exercises 4-3 and 4-4?

4-5. Compute s and the associated degrees of freedom for the following data set, consisting of five sets of triplicate measurements:

Set No.	Measured Values
1	20.5, 21.0, 20.6
2	20.9, 20.7, 20.9
3	20.7, 21.3, 20.9
4	20.8, 20.8, 20.5
5	21.0, 21.1, 20.6

a. Using the range.
b. Using the conventional formula, then pooling the five estimates.

4-6. A certain measurement process was considered to consist of five independent steps with individually estimated independent standard deviations as follows:

$$s_1 = 1.05, s_2 = 1.72, s_3 = 0.51, s_4 = 0.85, s_5 = 2.50$$

Estimate the overall standard deviation, s_T, for the measurement process.

4-7. While analyzing waste water for constituent A, the following sets of duplicate measurements were made to estimate the precision of measurement (all data in units of mg/L):

Set No.	X_f	X_s
1	20.7	21.0
2	21.1	20.9
3	21.0	21.1
4	23.0	23.0
5	18.3	17.8
6	19.9	20.0

a. Compute s on basis of differences, d.
b. Compute s on basis of range.
c. How many degrees of freedom are in each case?

4-8. Criticize the following statement taken from a report of analysis:

A sample of fuel oil was analyzed and found to contain 1.05% sulfur, with a standard deviation of 0.012%.

4-9. The standard deviation of a method of measurement was estimated on 5 occasions with the following results:

Occasion	s	n
1	0.75	10
2	1.45	5
3	1.06	7
4	2.00	7
5	1.25	12

a. Compute the pooled estimate of the standard deviation.
b. State the degrees of freedom upon which the value in a is based.

4-10. What is the standard deviation of a mean and how may it be estimated? Also, what is variance and how can it be estimated? Complete the following table, filling as many blanks as possible:

s	n	s^2	$s_{\bar{x}}$
1.22	10	_____	_____
_____	_____	3.45	.929
_____	7	_____	5.29
2.44	_____	_____	5.29
1.55	_____	_____	_____

REFERENCES

[1]   Natrella, M.G. *Experimental Statistics* NBS Handbook 91, (Gaithersburg, MD: National Institute of Standards and Technology, 1963).

[2]   Taylor, J.K. *Quality Assurance of Chemical Measurements* (Chelsea, MI: Lewis Publishers, Inc. 1987).

CHAPTER 5

Data Analysis Techniques

Many of the decisions related to the interpretation of scientific and engineering data can be made using a relatively small number of statistical techniques that are both simple and easy to use. This chapter describes the techniques that the author has found to be useful on many occasions for data analysis and data based decisions.

INTRODUCTION

The basic statistical data of Chapter 4 can be used to make a number of rather simple but important calculations that provide the basis for statistical decision making. The statistician calls this hypothesis testing and ordinarily the null hypothesis is tested. Essentially, the null hypothesis is that there is not a significant difference between two results. It will be seen that differences may have to be quite large in some instances before they are statistically significant, especially in the case of small data sets of high variability. Statistics will not say whether or not an apparent difference is real, but will only give the probability that it could have been as large as it is by chance alone. Often, the answer will be that there is no reason to believe a difference exists other than due to a chance occurrence, based on the statistical evidence available. Remember that this is not saying that there is no difference but that the evidence presented is insufficient to support the belief that the difference is not more than a random effect.

Based on natural variability, there will be some level of probability – a level chosen arbitrarily – where any chosen difference is significant. Indeed, one can determine what that probability is in any specific situation. This is referred to as the p value approach and is discussed in Chapter 9. The alternative is referred to as the criticial value approach, in which one sets the probability, such as a 95% confidence level, for a decision. The critical value approach is discussed more fully in the succeeding sections.

Before considering the statistical techniques to be used, one should think about the arithmetic involved. The question is frequently asked, "How many figures should be carried in calculations?" If one does not carry enough, rounding errors can influence some decisions in certain cases. For example, 1.5 and 2.5 round off to a value of 2 by commonly used rounding rules. It is evident that quite a different conclusion could be reached in some cases as the result of such a rounding. While this may be considered an extreme case, the principle is valid in every case. Modern computer procedures automatically round and avoid such problems, if they are properly programmed. The computer program should be checked if there is any doubt of what is happening in a particular case.

The rules for rounding were reviewed in Chapter 4. The computational rules relating to statistical calculations that the author has been taught by many mentors are as follows:

- Calculate all means to *at least* one more significant figure than is in the data set.
- Calculate standard deviation estimates to *at least* two significant figures.
- Calculate confidence intervals and other statistical quantities to two significant figures.
- Report means consistent with their confidence intervals

ONE SAMPLE TOPICS

Means

Many common statistical calculations involve the sample mean. One possible question involves how close the measured value, $\overline{X}$, is to the population mean, μ. This is impossible to say, but one can calculate a confidence interval within which the population mean is expected to lie, with a given probability.

Confidence Intervals For One Sample

The statistical relationship to be used for confidence intervals around the mean depends on whether one knows the population standard deviation, σ, or only its estimate, s.

In the case of known standard deviation, the confidence interval around the mean (m_l, m_u) is given by

$$\begin{array}{ll} \text{Confidence interval} & m_l = \overline{X} - z_{1-\alpha/2}\, \sigma/\sqrt{n} \\[4pt] \text{for } \mu, \text{known } \sigma & m_u = \overline{X} + z_{1-\alpha/2}\, \sigma/\sqrt{n} \end{array}$$

where $z_{1-\alpha/2}$ = probability factor for a two-sided hypothesis (consult Table A.2 in Appendix A for values for Z)

σ =   known value for the standard deviation

n   =   number of measurements used to calculate the mean

For a one-sided hypothesis, use $z_{1-\alpha}$.

Usually, one will not know σ but rather will have to estimate s from a set of data. In the case of unknown standard deviation, the confidence interval around the mean (m_l, m_u) is given by

$$\begin{array}{ll} \text{Confidence interval} & m_l = \overline{X} - t_{1-\alpha/2,k}\, s/\sqrt{n} \\[4pt] \text{for } \mu, \text{unknown } \sigma & m_u = \overline{X} + t_{1-\alpha/2,k}\, s/\sqrt{n} \end{array}$$

where   s = estimate of the standard deviation

$t_{1-\alpha/2,k}$ = "Student's t", the value of which depends on the probability level chosen for the interval and on the degrees of freedom upon which s is based. (See Table A.3 in Appendix A for numerical values)

n = number of measurements used to calculate the mean

The t table has columns labeled across the top to indicate the probability level desired. The far left-hand column lists the degrees of freedom.

As an illustration of its use, the value for t for a confidence level of 95%, and an s based on 7 degrees of freedom is 2.365. It will be noted that t is large when df is small and decreases to the Z value at df = ∞. It can be seen how wrong one would be if one calculated a "2-sigma" value for a 95% confidence interval, using an s based on a few degrees of freedom.

For a one-sided hypothesis, use $t_{1-\alpha,k}$.

Example To emphasize the use of the normal and t distributions, the reader is asked to complete the following table.

| | | 95% Value | | |
n	df	Z	t	t/Z
2	‾‾‾	‾‾‾	‾‾‾	6.353
4	‾‾‾	‾‾‾	‾‾‾	‾‾‾
6	5	2.0	2.571	1.285
8	‾‾‾	‾‾‾	‾‾‾	‾‾‾
10	‾‾‾	‾‾‾	‾‾‾	‾‾‾
15	‾‾‾	‾‾‾	‾‾‾	1.072
20	‾‾‾	2.0	‾‾‾	‾‾‾

Thus, for six measurements, a confidence interval based on the Z value would be ±28.5% too narrow. The reader may wish to plot t/Z with respect to n to get a better picture of how large an error is introduced by calculating "Z sigma" limits when "t s" limits are the correct ones.

The value for t is always larger than Z to take into account the fact that the standard deviation is not known with high confidence, when based on a limited amount of data. It will be seen what is the confidence of an estimate of σ later in the chapter. Another way to look at the situation is that t is a safety factor to prevent one from defining too narrowly the confidence interval, based on a small amount of data.

Concerning degrees of freedom, if $\overline{X}$ and s are based on the same data set, df = n − 1. However, if s is based on additional evidence, such as a system under statistical control (perhaps even judged by control chart evidence) [1], then the degrees of freedom on which the estimate of s is based may be used for determining the value for t. In such a case, one can calculate a confidence interval for even a single measurement.

Example

n = 7, s = 1.2	$\overline{X}$ = 12.52
95% C.I.	$12.52 \pm 2.447 \times 1.2 \div \sqrt{7} = 12.52 \pm 1.11$
99% C.I.	$12.52 \pm 3.707 \times 1.2 \div \sqrt{7} = 12.52 \pm 1.68$

n = 2, s = 1.2	$\overline{X}$ = 12.52
95% C.I.	$12.52 \pm 12.706 \times 1.2 \div \sqrt{2} = 12.52 \pm 10.8$

n = 1	s = 1.2 df = 20 $\overline{X}$ = 12.52
95% C.I.	$12.52 \pm 2.086 \times 1.2 \div \sqrt{1} = 12.52 \pm 2.50$

Scan Table A.3 in Appendix 3 horizontally and then vertically and see how the value of t varies. This should help one answer such questions as how a confidence interval could be narrowed by more measurements with increased effort, of course, or how much work would be saved if one were willing to take a bigger risk, that is, to have less confidence in a decision.

One other comment about confidence intervals is of importance. No matter what is the value of the standard deviation, the interval shrinks to smaller values as n increases and becomes zero for an infinite number of measurements. Remember that the interval decreases in inverse proportion to the square root of n, so it is not practical to approach zero by this way. Also, the interval decreases directly as s is decreased, which ordinarily results from increased precision due to improved quality control or even as the result of research investigations.

Before leaving this subject, the meaning of a confidence interval should be clarified. It means simply this. If one were to make a large number of sets of measurements, like the one just made, and calculate the confidence interval each time, one would expect (on the average) such intervals to include the mean the selected percentage of times, say 95 out of each 100 times for a 95% confidence interval. However one would not know on any particular occasion whether the interval calculated did or did not include the mean. For a 99% interval, the odds are 99 in 100 that it does, while a 99.9% interval increases the odds to 999 in 1000, but the intervals will increase in size as the odds are increased. A 95% interval is reasonably conservative and acceptable to most chemists and data users. Remember – *one must always state what the interval is.* A ± value means nothing without stating what it is. *Do not* calculate "2-sigma" and "3-sigma" limits unless it is really believed that sigma is *known!*

Just to show what confidence limits mean, Figure 5.1 which is based on random drawings from a normal population has been included.

Ninety percent confidence intervals are drawn so that the reader can count to verify the concept if she/he wishes to do so. The means are at the center of the intervals, of course. It will be seen that 90 of the intervals do include the true value of 50,000. The reader might like to select intervals from the figure, using the random number Table A.11 in Appendix A (when you come to that part of the book) to see how often an interval does and does not include the mean.

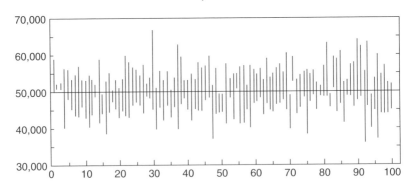

Figure 5.1. 90% confidence intervals.

Does a Mean Differ Significantly from a Measured or Specified Value?

The analyst is frequently confronted with the decision on whether a measured mean which appears to be different is really different from some specified value. The standard or specified value may be, for example

- A certified value for a reference material. If the measured value differs significantly, it could indicate that the measurement system is biased.
- A specified value that a product must meet. If the difference is real, the product does not meet specification and should be rejected.

Judging the possible violation of an environmental regulation is another of the many decisions of this type. In fact, analyses are often made in order to answer such questions and close decisions are typical situations. It is suggested that the reader look once again at Figure 3.1 in Chapter 3 in this regard.

The statistical calculations required are simple and have been already covered. As described, one calculates a confidence interval for the result of measurement and compares it with the specified value. If the specified value falls within these limits, there is no reason to believe, within the probability level chosen, that the measured value differs from the specified value.

Example Much fuel oil, to be acceptable, may contain no more than 1.00% sulfur. Five measurements of it were made and the results expressed as % S were 0.99, 1.01, 1.00, 0.99, 1.03. Does the oil meet the specification?

From the above data one would calculate

$$\overline{X} = 1.004 \quad s = 0.0167 \quad n = 5 \quad 95\%\text{C.I.} = 1.004 \pm .021$$

Since the specified value of 1.00 is within the confidence interval given by (0.983,1.025), there is no reason to believe that the specification is not met, based on the measurements.

Example Suppose that a more precise method were used and the following results were obtained: 0.998, 1.008, 1.001, 0.995, 1.003. What would be the conclusion? From the above data one would calculate

$$\overline{X} = 1.001 \quad s = 0.0049 \quad n = 5 \quad 95\% \text{ C.I. } \overline{X} = 1.001 \pm .006$$

Since the confidence limits (0.995,1.007) include 1.00, there is still no reason to believe that the specification has been exceeded.

The key words above are "no reason to believe". In other words, some of the measured values did and could have exceeded the specified limits on the basis of chance and the variability of the measurement process. Thus, based on 95% confidence intervals there is no reason to believe that 12 ± 5 exceeds 10.0 while there is reason to believe that 10.2 ± 0.1 does exceed the value. If a difference of 0.2 units were important, then the first measurement process is inadequate to make the judgment of compliance or noncompliance. Such matters always need to be considered when making measurements and interpreting data. The important question of the magnitude of the standard deviation that is permissible to detect a given difference is discussed in Chapter 9.

This leads to a further consideration of 'significance'. One needs to distinguish between statistical significance and "real world" significance. A result could have no statistical significance but could be disastrous if real. This might be because the data were insufficient, for example, so that the power of the statistical test was inadequate. (See Chapter 9 for a further discussion of this subject.) On the other hand, a difference in two measurements could be indicated as statistically significant, yet would be of no practical importance. Remember that statistics are for guidance as mentioned in Chapter 3. However, do not use this statement to excuse poor data!

MINITAB Example

MINITAB offers features for calculation of confidence intervals around the mean. The calculation is illustrated using the MINITAB furnace temperature dataset, 'furntemp.mtw', which provides measurements of furnace temperatures under varying conditions of plant, shift, operator, and batch. For this example, only the temperature information is used. The first 10 observations of the furnace temperature dataset are shown in Table 5.1:

Table 5.1.   Furnace Temperature Dataset from MINITAB

Furnace Temperature
477
472
481
478
475
474
472
475
468
482

Taken from MINITAB

Assume that the plant operators know that the standard deviation of the temperature is 3, and they are interested in calculating a 90% confidence interval around the mean temperature. The mean temperature should be 475 degrees for the optimum product. The following MINITAB menu commands are used to calculate a 90% confidence interval around the mean furnace temperature using the entire data set (n=192):

Under Stat, Basic Statistics, 1 Sample Z
- double click on C1 temp to add it to the **Samples in Columns** field
- enter 3 in the field for **Standard Deviation**, enter 475 in the field for **Test Mean**
- under **Options**, enter 90 in the field for **Confidence Interval**, click OK, click OK

The results are shown below:

MINITAB Output

```
Results for: Furntemp.MTW

One-Sample Z: Temp

Test of mu = 475 vs mu not = 475
The assumed sigma = 3

Variable          N      Mean     StDev    SE Mean
Temp            192   474.880     4.775      0.217

Variable              90.0% CI             Z      P
Temp        ( 474.524, 475.236)       -0.55  0.580
```

Taken from MINITAB

The 90% confidence interval (shown in bold) is (474.5, 475.2). If the plant operators had not known the standard deviation of the temperature, a t distribution based confidence interval should be calculated:

Under Stat, Basic Statistics, 1 Sample t
- double click on C1 temp to add it to the **Samples in Columns** field
- enter 475 in the field for **Test Mean**
- under **Options**, enter 90 in the field for **Confidence Interval**, click OK, click OK

The output is as follows:

MINITAB Output

```
One-Sample T: Temp

Test of mu = 475 vs mu not = 475

Variable          N      Mean     StDev    SE Mean
Temp            192   474.880     4.775      0.345

Variable              90.0% CI             T      P
Temp        ( 474.311, 475.450)       -0.35  0.728
```

Taken from MINITAB

The 90% confidence interval (shown in bold above) is (474.3, 475.5). Note how similar the Z and t confidence intervals are for this example. This is due to the large number of temperature readings available (n=192). Since both the confidence intervals include 475 degrees, the plant operators do not have any reason to believe that the process is inadequate.

Standard Deviations

As was done in the case of means, one may estimate and make inferences regarding the population standard deviation.

Confidence Intervals For One Sample

It is possible to calculate a confidence interval around an estimate of a standard deviation. It has been said earlier that each time one estimates this parameter, one expects to get a somewhat different value. How do they relate to the population standard deviation, σ? Again one cannot exactly answer such a question but can calculate an interval that will include σ with a given probability, that is, for a specified percentage of times. The interval that is appropriate is not symmetrical as it was in the case of the mean because the distribution is not normal. This is because one tends to underestimate standard deviations when small numbers of data are used. The value could be somewhat smaller but it could also be considerably larger than the experimentally determined one. This can be looked upon in the following way. The standard deviation is a measure of variability and one seldom sees the full range of variability of a population when only a few individuals are observed.

The expression to use in calculating the confidence interval (σ_l, σ_u) for a standard deviation is

$$\boxed{\text{Confidence Interval for } \sigma \quad \begin{aligned} \sigma_l &= B_L \times s \\ \sigma_u &= B_U \times s \end{aligned}}$$

where values for B_L and B_U depend on the confidence level desired and the degrees of freedom upon which s is based. Values for B_L and B_U are found in statistical tables. The symbols L and U refer to lower and upper limits, respectively. An abridged table (Table A.4) is given in Appendix A.

Example Suppose that s is based on 9 degrees of freedom (a set of 10 measurements) and a 95% confidence is required for the interval. The value for B_L is 0.666 and the value for B_U is 1.75. Suppose now that the estimate of the standard deviation was s = 1.2. Then the confidence interval is (0.8,2.10). If a confidence of 99% were desired, the interval would be (0.72,2.60). The reader is asked to check this calculation to be sure that the principle involved is understood.

Example To illustrate the idea further, suppose that one made two measurements with results of 12.2 and 12.8. What would be the 95% confidence interval for the standard deviation estimate? The answer is (0.15,7.6)! Astonishing? Check the author's calculations!

Again, it is suggested that the reader look at the table very thoughtfully. It will give one a new insight into variability. One should conclude, if nothing else, that much work is required whenever the extent of variability and its influence are of concern. Estimated means converge more rapidly toward the population value than estimates of standard deviations. Fortunately, any given estimate of s is the most probable value and the uncertainty of s values is compensated for by the Student's t when making calculations of confidence intervals for means. Looking ahead, it will be seen that the large uncertainty in s values makes it difficult to decide whether an apparent difference of two such estimates is significant. The F test will give one guidance in this matter, as will be seen later.

Does a Standard Deviation Differ Significantly from a Measured or Specified Value?

A serious analyst is always concerned about the precision of her/his methodology and the data resulting from its use. One such concern is how the precision attained compares with that obtained by others. The simplest question of this nature is how the precision compares to a so-called standard or specified value, such as one quoted for a standard method or perhaps in the Federal Register. In fact, a regulatory body or even a prospective client may require that a stated standard deviation be achieved. How can a statistically supported judgment be made on matters of this kind? The statistical technique useful for the above purpose is simple and easy to apply. All one needs to do is to select the confidence level for the decision, and then compute the statistical confidence interval for the estimated standard deviation.

The information needed is the value for s and the degrees of freedom on which the estimate is based. It will be remembered that the interval is given by

$$B_L \times s \text{ to } B_U \times s$$

If the standard (or required/specified) value lies within these limits, there is no reason to believe, at the confidence level selected, that the experimental value differs significantly from it. If the experimental lower limit is above the standard value, one believes that the value for s is high. If the standard value is above the higher limit, it may be concluded that the experimental value is smaller than the standard value, that is, that the measurement process is more precise than the one that produced the "standard" data.

Example   The facts are as follows:

Standard value s (standard)	= 1.8 mg/L
Experimental value	= 2.0 mg/L df = 6
Decision level	95% confidence

From Table A.4 in Appendix A, one finds $B_L = 0.614$; $B_U = 2.05$. The 95% confidence interval is thus

$$(0.614 \times 2.0, 2.05 \times 2.0) \quad \text{or} \quad (1.23, 4.10).$$

It would be concluded that there is no reason to believe at the 95% level of confidence that the experimental value differs from the standard value.

The same concept can be applied to limits of detection or to sensitivity, which involve simple multiples of a standard deviation [1]. In such cases, one should convert the parameter in question to a standard deviation by dividing by the appropriate factor, then test for significance of the observed value with respect to it.

For example, The American Chemical Society's Committee on Environmental Improvement [2] defines the Method Limit of Detection (MDL) as 3 s_0, where s_0 represents the standard deviation at the "zero" or lowest level of measurement of the methodology. If a MDL of 1 mg/L is specified, the value for s_0 that a laboratory must attain is 0.33 mg/L. Evidence that such an s_0 value is not only attained but consistently maintained is the best evidence for the attained/maintained MDL of a laboratory.

Minitab Example

The graphical summary feature in MINITAB offers calculations of confidence intervals for standard deviations as well as means. In addition, a normal curve is superimposed on a histogram of the data to help assess normality. The following menu commands produce a graphical summary for the furnace temperature data:

Under Stat, Basic Statistics, Graphical Summary
• double click on C1 temp to add it to the **Variables** field, click OK

For the furnace temperature dataset, the graphical output is shown in Figure 5.2.

The confidence interval around the standard deviation of the temperature is (4.34, 5.31).

Statistical Tolerance Intervals

In the previous sections, confidence intervals for means were calculated and it was remarked that the width of the interval could be reduced (at least in principle) to very low values and it could even approach zero for enough repetitive measurements. Of course, this would mean that the value for the mean was known "perfectly".

There are times when the bounds for the population are desired. Ordinarily, this will be some definite percentage of the population, say the bounds within which 95 or 99% of the population is expected to lie. To make such a prediction from a set of data, one would need to know the mean and the standard deviation. In the case of population parameters, that is, known values for μ and σ, the calculation is easy and exact. The expression to be used is

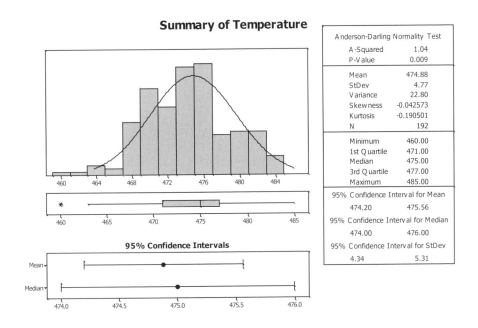

Figure 5.2. Graphical summary including confidence interval for standard deviation.

$$\text{Tolerance interval} = \mu \pm Z\sigma$$
$$\text{for known } \mu, \sigma$$

where Z is selected according to the percentage of the population to be included, using a table such as Table A.2 in Appendix A. Thus, 68% of the population lies within ±1σ of the mean and 90% lies within ±1.645 σ. It is left to the reader to decide what percentage of the population lies within ±3 σ of the mean. The reader is asked to calculate where 50% of the population would lie, centered about the mean.

In the real world, one often has to base such predictions on small amounts of data, which means that $\overline{X}$ and s may be all that are available. In this case, one must use a somewhat different expression to calculate the limits, which will be called statistical tolerance limits, namely

$$\boxed{\begin{array}{l} \text{Statistical Tolerance interval} = \overline{X} \pm ks \\ \text{for unknown } \mu, \sigma \end{array}}$$

where k is the statistical tolerance factor. The values for k, found in statistical tables (Table A.5 in Appendix A, for example) depend on three things: (1) the proportion, p, of the population that one wishes to include, (2) the probability, τ, of inclusion in the interval that is calculated, and (3) the number of degrees of freedom on which s is based. For example, one could calculate an interval for which one had a 95% confidence that it would contain 95% of a given population.

Example   A drug manufacturer produces tablets, each intended to contain a specified amount of medicine. On the average, the tablets may conform to the specification. However, the individual tablets will vary and their standard deviation about the mean can be estimated. The manufacturer would like to know the limits within which some percentage, say 99%, of the tablets lie, and he would like to have, shall we say, a 95% confidence for this estimate. Obviously, he could not have a 100% confidence because of the uncertainty of the data. Suppose 25 tablets were tested and the average content was 100 mg with a standard deviation estimate of 1.1 mg. This information may be used to calculate the 95%/99% tolerance interval. From Table A.5 in Appendix A, one finds for n = 25, τ = 95, p = 99, k = 3.46. From this, one could calculate that $100 \pm 3.46 \times 1.1 = 100 \pm 3.8$ mg. In other words, the manufacturer would have a 95% confidence that 99% of the tablets contain between 96.2 and 103.8 mg of the active ingredient.

To carry this discussion a little further, suppose that the tests were expensive and the manufacturer did not want to test so many tablets. What would have been the conclusion if 10 tablets had been tested with the same values for s and $\overline{X}$? The value for k in this case is k = 4.43 and the tolerance interval is 100 ± 4.9 mg. It will be left to the reader to calculate what would have been the conclusion if he had been so foolish as to try to get by with testing only a pair of tablets, assuming the same values for s and $\overline{X}$.

When statistically analyzing data, one should always ask whether the mean or the population bounds is the important issue. If it is the population bounds, then the number of samples to be analyzed may need to be much larger than in the case of a mean. However, now one is getting into sampling problems which, though related, soon get beyond the scope of the present book.

Table 5.2 is included to provide further perspective on the relation of confidence intervals and tolerance intervals. It contains the numerical factors by which a known or estimated standard deviation should be multiplied to obtain 95% confidence intervals or 95% tolerance intervals with 95% confidence, respectively. One will note that confidence intervals narrow much more rapidly than tolerance intervals. It is further evident that confidence intervals based on known values for the standard deviation also decrease more rapidly than those based on estimates from the sample used to compute the value of the mean.

The techniques described above for estimating statistical tolerance intervals are based on the assumption that the data of concern are normally distributed. If this is not true, or the distribution is not known, one can still calculate statistical tolerance intervals using what is called a distribution-free approach, provided sufficient data are available. The individual data points are ranked, serially, from the smallest to the largest value. Tables are available that contain values for r and s such that one may assert with a confidence of at least τ that a designated percentage of the population will lie between the rth smallest and the sth largest of a randomly selected set of n individuals from the population. The minimum value for n is 50.

Table 5.2. Comparison of Confidence and Tolerance Interval Factors

n	Standard Deviation Known		Standard Deviation Estimated	
	95% C.I.	95% T.I.	95% C.I.	95% T.I.
2	1.385	1.96	8.984	37.674
3	1.132	1.96	2.484	9.916
4	.980	1.96	1.591	6.370
5	.877	1.96	1.241	5.079
6	.800	1.96	1.050	4.414
7	.741	1.96	.925	4.007
8	.693	1.96	.836	3.732
9	.653	1.96	.769	3.532
10	.620	1.96	.715	3.379
20	.438	1.96	.466	2.752
50	.277	1.96	.284	2.379
100	.196	1.96	.199	2.233
∞	0	1.96	0	1.96

Example For a set of 100 data points, there is a 95% confidence that 90% of the population will have values within the third smallest and the second largest value. Reference 3 contains Table A-30 for use in calculating two-sided distribution-free tolerance intervals and Table A-31 for calculating one-sided intervals.

Combining Confidence Intervals and Tolerance Intervals

The procedure described in the previous section is for the situation where all of the variance is due to the variability of the material. In other words, it is assumed that the variance of the measurement process is negligible with respect to that of the material measured. A good rule-of-thumb is that when the standard deviation of either contributor – material or measurement process – is no more than one third of the other, the smaller one may be considered to be negligible for all practical purposes.

When the variances of both contributors are significant one needs to evaluate each, independently, using the analysis of variance method described in Chapter 4. Another approach is to evaluate the variance of the measurement process by measurement of similar but homogeneous samples [1].

When s_m and s_a are known, one can estimate the confidence interval for the mean by the expression given in the section Means, Confidence Intervals for One Sample and the statistical tolerance interval for the population of samples of the material from that given in the previous section. The question of how to combine such intervals is answered differently by different statisticians. The author contends that the most conservative way to estimate the total interval is to add the confidence interval to the tolerance interval. An example of such a calculation follows.

Example Measurements were made on 10 samples of waste water for its cyanide content, with the following results:

$$\overline{X} = 1.35 \, mg/L$$
$$s_a = 0.062 \, mg/L$$
$$s_m = 0.15 \, mg/L$$

The half widths of the respective confidence and tolerances are calculated as follows:

$$95\% \, C.I. \text{ for mean} = 2.26 \times .062 / \sqrt{10} = 1.35 \pm 0.044 \, mg/L$$
$$95\%, 95\% \, T.I. = 3.38 \times 0.15 = 1.35 \pm 0.507 \, mg/L$$

When the uncertainty of the mean is considered, material variability is considered to be

$$1.35 \pm (.507 + .044) = 1.35 \pm 0.55 \, mg/L$$

The above concept is illustrated by Figure 5.3. The spread of the values around the central line is represented by the confidence interval for the mean. The satellite distributions are due to the variability of the material. It should be evident that, as the central distribution is narrowed either by additional measurements or by use of a more precise method (smaller s_a), all of the variability will be due to sample inhomogeneity. Also, one can visualize the case where the central distribution is overwhelmingly larger than the satellites ($s_a \gg s_m$). The satellite distributions will become negligible, and all variability is due to measurement. When the respective variances are each significant, then both contributions must be considered in putting realistic limits on the estimates of the composition of the material analyzed.

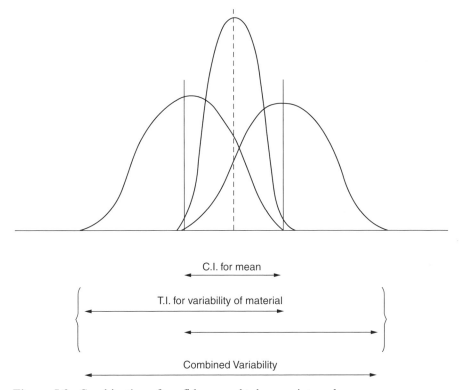

Figure 5.3. Combination of confidence and tolerance intervals.

TWO SAMPLE TOPICS

Means

Do Two Means Differ Significantly?

One is often confronted with the question of whether two means really differ. Two measured values or two averages, X_A and X_B, will seldom be identical, but is the difference significant? Statistical guidance will be needed in making such decisions in most cases, especially when the two values approach each other. Typical situations include

• Values obtained by the same analyst on two occasions, A and B
• Values reported for measurements on the same material by two analysts, A and B, in the same or different laboratories
• Values obtained using two diferent measurement techniques, technique A and technique B

In order to make a statistical decision, one must know the respective standard deviation estimates and the number of measurements used to calculate each mean. Thus, one needs to know

$$\overline{X}_A, s_A, n_A$$
$$\overline{X}_B, s_B, n_B$$
$$\Delta = |\overline{X}_B - \overline{X}_A|$$

One of the following situations is possible:

1.   Case I. The values for s_A and s_B do not differ significantly
2.   Case II. The values for s_A and s_B differ significantly

It will be seen that the F test can be used when deciding whether the values for s differ significantly. Sometimes such a test is not needed because the standard deviations are known to be different because the populations themselves are different (see Chapter 4). The statistical calculations to be made are the following:

CASE I.   The standard deviations do not differ significantly.

Step 1.   Pool the s values to obtain s_p, a better estimate of the standard deviation. (This is done as already described in Chapter 4). df = $n_A + n_B - 2$
Step 2.   Calculate the variance of $\overline{X}_A$ and $\overline{X}_B$

$$V_A = \frac{s_p^2}{n_A} \qquad\qquad V_B = \frac{s_p^2}{n_B}$$

Step 3.   Choose the probability level (i.e., the significance level) for the decision
Step 4.   Calculate the uncertainty of the difference of the two means at the proba-bility level chosen $U_\Delta = t\sqrt{(V_A + V_B)}$ using a value for t corresponding to df = $n_A + n_B - 2$ and the probability level chosen
Step 5.   If Δ is greater than U_Δ, conclude that the means differ; if Δ is not greater than U_Δ, there is no reason to believe that the means differ

CASE II.   The standard deviations differ significantly.

Step 1. Calculate the variances for the respective means

$$V_A = \frac{s_A^2}{n_A} \qquad V_B = \frac{s_B^2}{n_B}$$

Step 2. Choose the probability level for the decision
Step 3. Calculate the *effective number of degrees of freedom*, f, to be used for selecting t when calculating U_Δ. This is necessary because adding df_A and df_B is like adding apples and oranges!

$$f = \frac{(V_A + V_B)^2}{\dfrac{V_A^2}{n_A - 1} + \dfrac{V_B^2}{n_B - 1}}$$

Step 4. Calculate the uncertainty of $U_\Delta = t^* \sqrt{(V_A + V_B)}$ where t^* is based on f and the chosen probability level.
Step 5. If U_Δ is greater than Δ, there is no reason to believe that the means differ.

Case II may look a little complicated, on first inspection, but it is really quite simple. The author always refers to the formulas, especially for Case II in the calculation of f, but he finds it easy to remember the rest. The reader will, too, if she or he makes such calculations frequently.

Example Consider Case I. The two sets of data are the following:

$$\overline{X}_A = 50\,mg/L \quad s_A = 2\,mg/L \quad n_A = 5$$
$$\overline{X}_B = 45\,mg/L \quad s_B = 1.5\,mg/L \quad n_B = 6 \qquad \Delta = 5\,mg/L$$

Step 1.
$$s_p = \sqrt{\left(\frac{2^2 \times 4 + 1.5^2 \times 5}{4 + 5}\right)}$$
$$s_p = \sqrt{3.028} = 1.74 \quad df = 9$$

Step 2.
$$V_A = \frac{3.028}{5} = 0.6056 \qquad V_B = \frac{3.028}{6} = 0.5047$$

Step 3. Decision to be made with 95% confidence
Step 4. $U_\Delta = 2.262\sqrt{(0.6056 + 0.5047)} = 2.38$
Step 5. Since Δ exceeds U_Δ, conclude that the means differ significantly at the 95% level of confidence.

The same example will be considered, but this time it will be assumed that the standard deviations do indeed differ. This could be because the data came from two different laboratories (see Chapter 4). Thus Case II applies.

Step 1. $V_A = 2^2 / 5 = 0.800$ $V_B = 1.5^2 / 6 = 0.375$
Step 2.   Decision to be made at 95% confidence level

Step 3. $$f = \frac{(.800 + .375)^2}{\dfrac{.800^2}{4} + \dfrac{.375^2}{5}} = 7.34 = 7$$

Step 4. $U_A = 2.365\sqrt{(.800 + .375)} = 2.6$
Step 5.   Since Δ exceeds U_A, conclude that the means differ at the 95% level of confidence.

The two examples above resulted in the same conclusions, regardless of whether Case I or Case II strategies were used. This certainly cannot be generalized. If in doubt, it is recommended that the Case II strategy be used, which is the more conservative approach. The general case for intercomparison of several means is discussed in Chapter 9.

MINITAB Example

Consider the MINITAB dataset 'acid.mtw', which shows a chemistry class' titration results from 2 acids. The first few observations are shown in Table 5.3.

Table 5.3. Acid Dataset from MINITAB

Acid 1	Acid 2
0.123	0.109
0.109	0.111
0.110	0.110

Taken from MINITAB

It is of interest to the class to examine, with 95% certainty, whether the mean titration results differ. The data provides 124 titration results for acid 1 and 37 titration results for acid 2. The MINITAB menu commands are as follows:

Under Stat, Basic Statistics, 2 Sample t
• select **Samples in Different Columns**, double click on Acid1 to add it to the **First** field and Acid2 to add it to the **Second** field, click OK

The output is shown below.

MINITAB Output

```
Results for: Acid.MTW

Two-Sample T-Test and CI: Acid1, Acid2
Two-sample T for Acid1 vs Acid2

          N       Mean      StDev     SE Mean
Acid1    124     0.11006    0.00454   0.00041
Acid2     37     0.10965    0.00184   0.00030

Difference = mu Acid1 - mu Acid2
Estimate for difference:   0.000416

95% CI for difference: (-0.000589, 0.001421)

T-Test of difference = 0 (vs not =): T-Value = 0.82   P-Value = 0.415
DF = 145
```

Taken from MINITAB

The 95% confidence interval (shown in bold) around the difference in means includes 0, indicating with 95% certainty that the 2 acids do not significantly differ from each other. Note that the 'assume equal variances' option was not selected due to the large difference in sample size between the acids. This could have been investigated using statistical tests for differences in variances, which are discussed in the next section.

Standard Deviations

Do Two Standard Deviations Differ Significantly?

The analyst is often concerned about the significance of apparent differences between two estimates of a standard deviation. Has the measurement precision changed for the better or the worse, or is one method of measurement more precise than another, or is the precision of one laboratory significantly different from that of another laboratory, are examples of such questions that may require decisions.

The statistical technique used when making such decisions is called the F test. Calling the two situations A and B, respectively, one needs to know s_A and s_B and the degrees of freedom on which each is based. Then one calculates the ratio

$$F = \frac{s_{larger}^2}{s_{smaller}^2}$$

This ratio will always be equal to or greater than 1. The value for F so calculated is compared with the critical value of F_c, found in statistical tables, such as Table A.6 of Appendix A. The value for F_c depends on the confidence level desired for the decision, and the degrees of freedom for the numerator and for the denominator. Table A.6 in Appendix A is for the 95% level of confidence for the decision. Values of F for other confidence levels will be found in statistical compilations such as Reference 3. This F test is two-tailed in that it really only answers the question of whether the standard deviations are different.

Example Assume that an analyst used a certain method to determine the amount of lead in a water sample and estimated s = 2.5 µg/L based on 7 measurements. At a later date, she analyzed another water sample for its lead content, using the same method, and obtained s = 3.3 µg/L based on 10 measurements. Were both sets of measurements equally precise with 95% confidence?

Calculate

$$F = \frac{3.3^2}{2.5^2} = 1.74$$

df = 9 for numerator, df = 6 for denominator; for 95% confidence, F_c = 5.52 (by interpolation).

Since F does not exceed F_c, conclude that, at the 95% level of confidence, there is no reason to believe that the precisions differ.

Suppose someone asked how large would s have to have been on the second occasion, before it would have been significantly different from the first measurement? The answer could be based on the following calculation:

$$F_c = 5.52 = \frac{s^2}{2.5^2}$$
$$s^2 = 34.5 \quad s = 5.9$$

It is left to the reader to make such calculations as: How small would the first estimate of s have to have been before one would conclude that the second estimate of 3.3 µg/L was larger than it?

How many measurements on each occasion would have been necessary to conclude that the two numerical values cited were different?

One might be asked the question, "Why are so many measurements required to decide that an apparent difference of 30% in the two estimates of s was significant?" A referral to the section Standard Deviations, Confidence Intervals for One Sample will emphasize how large the confidence intervals are for standard deviation estimates based on small data sets. When seen in this light, the answer will be intuitively clear.

Because the F test is so insensitive to small differences in the values for s, one may need to make judgments on the facts, i.e., on a physical basis, and not necessarily to rely on statistics in some situations. Thus, if a measurement process has been altered for some reason, perhaps one should assume that the precision will not necessarily remain the same, and use a newly determined value for s, regardless of the outcome of an F test. Similarly, one may change a control limit on such a basis, although one could not say with a reasonable confidence that the estimated values were statistically significantly different [1].

MINITAB Example

Recall the earlier acid titration dataset. Rather than assuming unequal variances for the calculation of the confidence interval around the difference in means (and hence not selecting a pooled estimate), one could have investigated this further by testing for differences between two variances in MINITAB:

Under Stat, Basic Statistics, 2 Variances
- select **Samples in Different Columns**, double click on Acid1 to add it to the **First** field and Acid2 to add it to the **Second** field, click OK

Which yields the following output:

MINITAB Output

```
Test for Equal Variances: Acid1 vs Acid2
Level1     Acid1
Level2     Acid2
ConfLvl    95.0000

Bonferroni confidence intervals for standard deviations
   Lower     Sigma      Upper     N   Factor Levels
3.97E-03  4.54E-03  5.30E-03    124   Acid1
1.46E-03  1.84E-03  2.49E-03     37   Acid2

F-Test (normal distribution)
Test Statistic: 6.071
P-Value        : 0.000

Levene's Test (any continuous distribution)
Test Statistic: 2.932
P-Value        : 0.089
```
Taken from MINITAB

Since $F=6.071> F_c=5.52$, the equal variance assumption cannot be selected. MINITAB also provides graphical output (shown in Figure 5.4) to support this conclusion.

Notice how the 2 confidence intervals around the standard deviations do not overlap. The boxplot indicates several outliers in the data for acid 1, which makes the estimate of sigma much larger than that of acid 2.

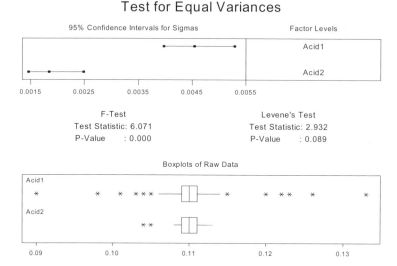

Figure 5.4. Tests for equal variances.

PROPAGATION OF ERROR IN A DERIVED OR CALCULATED VALUE

There are many experimental situations in which measured values in themselves are not the item of interest but rather the quantities derived or calculated from them. Thus, one may measure the mass of a liquid contained in a pycknometer to calculate its density. The measured volume (calibration of the pycknometer) is also involved. How do the variances of the two measured values influence the variance of the calculated value?

Consider the general case in which a measured value of x, m_x, and a measured value of y, m_y, are used to calculate m_w. These quantities are functionally related by

$$m_w = f(m_x, m_y)$$

If the function is known, one can partially differentiate it to evaluate the contributions of the variance of each variable to that of the calculated quantity. Ku [4] has done this for a number of simple functional relationships (Table 5.4).

It is assumed that the values of the measured quantities are independent, and no correlations exist. While the table relates to only a maximum of two variables, the same approach can be applied to three or more variables if applicable. Dr. Ku's paper should be consulted for more information on this subject.

Table 5.4. Propagation of Error Formulas for Some Simple Functions

Function Form	Approximate Formula for s_w^2
$M_w = Am_x + Bm_y$	$A^2 s_{\bar{x}}^2 + B^2 s_{\bar{y}}^2$
$M_w = m_x / m_y$	$(\bar{x}/\bar{y})^2 (s_{\bar{x}}^2/\bar{x}^2 + s_{\bar{y}}^2/\bar{y}^2)$
$M_w = 1/m_y$	$s_{\bar{y}}^2/\bar{y}^4$
$M_w = m_x \, m_y$	$(\bar{x}\bar{y})^2 (s_{\bar{x}}^2/\bar{x}^2 + s_{\bar{y}}^2/\bar{y}^2)$
$M_w = m_x^2$	$4\bar{x}^2 s_{\bar{x}}^2$
$M_w = \sqrt{m_x}$	$1/4(s_{\bar{x}}^2/\bar{x}^2)$
$M_w = \ln m_x$	$s_{\bar{x}}^2/\bar{x}^2$
$W = RSD = 100 \, (s/\bar{x})$	$\bar{w}^2/[2(n-1)]$ where n = number of values used to calculate $\bar{x}$

All variances are the variances of means. It is assumed that the values of x and y are independent and that the variances are small with respect to the measured values [4].

Example The following will illustrate the use of Table 5.4.

The density of a liquid was measured by pycknometry, with the following experimental results:

1. Ten independent measurements of mass of contents, m_x
2. Five independent calibration measurements of volume of pycknometer, from mass of contained water, m_y

$$\bar{x} = 9.23675\text{g}, \quad n = 10, \quad s_x = 0.00011, \quad s_{\bar{x}}^2 = 1.21 \times 10^{-9}$$

$$\bar{y} = 10.00155\text{mL}, \quad n = 5, \quad s_y = 0.000081, \quad s_{\bar{y}}^2 = 1.31 \times 10^{-9}$$

$$\text{density} = m_w = 9.23675/10.00155 = 0.92353\text{g/mL}$$

$$s_w^2 = (.92353)^2 (1.21 \times 10^{-9}/9.23675^2 + 1.31 \times 10^{-9}/10.00155^2)$$

$$s_w^2 = (.85291)(1.418 \times 10^{-11} + 1.310 \times 10^{-11}) = 2.33 \times 10^{-11}$$

$$s_w = 4.83 \times 10^{-6} \text{ g/mL}$$

Note: units for s and s^2 were omitted from the above tabulation

In addition to estimating a variance of a calculated result, the formulas in Table 5.4 may be used in the design of experiments to determine how many replicate measurements are needed to reduce the variance of a measured result to a desired level.

EXERCISES

5-1. Do the following arithmetic and report your answers to the proper number of significant figures:

$$(12.3 + 125.6 + 4.25 + 7.61) \times 1.2675 = \underline{\hspace{2cm}}$$
$$(14.32 \times 7.575) / (3.75 + 120.75) = \underline{\hspace{2cm}}$$
$$0.0000285 \times 1.00002567 = \underline{\hspace{2cm}}$$
$$12.345 + 10.000 + .1254 + 10670 = \underline{\hspace{2cm}}$$

Round off to 3 significant figures:

$$12.35 = \underline{\hspace{2cm}}$$
$$126.55 = \underline{\hspace{2cm}}$$
$$0.0325789 = \underline{\hspace{2cm}}$$
$$12.45 = \underline{\hspace{2cm}}$$

5-2. Explain what is meant by a statistical confidence interval and how it should be interpreted.

5-3. You reported the following information to a client: $\overline{X} = 20.3 \pm 1.8$, where the $\pm$ value is the 95,95 tolerance interval. The client asked for an explanation. What should you tell him?

5-4. Very careful measurements were made for the lead content on randomly selected samples of a proposed reference material with the following results, all expressed in mg/L. Use the software of your choice for each computation.

3.50, 3.57, 3.57, 3.53, 3.59, 3.49, 3.55, 3.57, 3.64

a. Make a plot to decide whether the distribution may be considered to be normal.
b. On the assumption that the material may be considered to be homogeneous, compute the 95% confidence interval for the mean.
c. On the assumption that the measurement process may be considered to have negligible variance, compute the 95, 95% tolerance interval and the 95, 99% tolerance intervals, respectively, for the composition of the samples.
d. If the measurement process were known to have a standard deviation of 0.030 mg/L, based on df = 30, compute the 95, 95% tolerance interval for the samples.

5-5. The standard deviation of a certain method was listed in a research paper as 0.12 mg/L. An analyst tried out the method on an appropriate sample and obtained the following data:

14.55, 14.47, 14.50, 14.82, 14.31, 14.61, 14.48

Was his precision significantly different from the reported value at the 95% confidence level? Answer this question using hand calculations and the software of your choice.

5-6. Laboratory A analyzed a sample of iron ore and obtained the following results, reported as wt % Fe:

65.97, 66.32, 66.15, 66.41

Laboratory B analyzed the same material and reported:

66.55, 66.85, 66.44, 66.60

Do the results of the two labs differ significantly at the 95% confidence level?

5-7. An analyst was investigating a new method of analysis for both its precision and possible bias. He decided to analyze a reference material that had a certified composition of 14.57 ± 0.02 for the above purpose. The following results were obtained:

14.55, 14.72, 14.66, 14.58, 14.70, 14.61, 14.56

a. What should he report for his attained standard deviation?
b. What can he say about possible bias of his measurements?

5-8. If Method A has a reported value for s = 1.75, based on df = 10, and method B has a reported value for s = 3.50, based on df = 6, can their precisions be considered to be significantly different at the 95% confidence level?

5-9. Complete the following table, filling in as many blanks, as possible:

$\overline{X}$	n	df	s_X	$s_{\overline{X}}$	$s_{\overline{X}}^2$	$t_{.975}$	$k_{95,95}$	95% CI	95,95%TI
2	___	___	___	___	___	5.08	___	.508	___
4	2	___	___	___	___	___	___	.45	___
6	___	7	.20	___	___	___	___	___	___
8	6	___	___	.102	___	___	___	___	___
10	___	___	.50	.158	___	___	___	___	___
15	20	___	.30	___	___	___	___	___	___
20	___	___	___	___	___	3.182	___	1.38	___

5-10. For a sample of a population with an assumed standard deviation of s = 1.00, plot the following on the same graph paper:

   a. The 95, 95% tolerance interval as a function of n for values of n from 1 to 100.
   b. The 95% confidence interval (ts/√n) for values of n from 1 to 100.
   c. The 95% confidence interval (zs/√n) as in b.
   d. Show the asymptotic value that each of the above approach as n → ∞.

5-11. Plot the 95% and the 99% confidence intervals for the standard deviation as a function of n over the range of 1 to 50.

5-12. Using the MINITAB dataset 'backpain.mtw' to answer the following questions.

   a. Calculate an 85% confidence interval around the cost of treatment for low back pain.

   b. Examine whether the variances around cost of treatment between genders are the same.

c. Calculate a 99% confidence interval to determine whether the average cost of treatment between genders is the same.

REFERENCES

[1] Taylor J.K., *Quality Assurance of Chemical Measurements* (Chelsea, MI: Lewis Publishers, Inc., 1987).

[2] American Chemical Society Committee Report, "Principles of Environmental Measurements," *Anal. Chem.,* 55: 2210-2218 (1983).

[3] Natrella M.G., *Experimental Statistics* NBS Handbook 91, (Gaithersburg, MD: National Institute of Standards and Technology).

[4] Ku H., *Precision Measurement and Calibration: Statistical Concepts and Procedures* NBS Spec. Publ. 300. Vol. 1, (Gaithersburg, MD: National Institute of Standards and Technology, 1969).

CHAPTER 6

Managing Sets of Data

The management of simple data sets is an important operation for many scientists and engineers. Data should be looked at, critically, for errors and consistency. Questions on whether all members of a set are peers or whether outliers are present need to be answered in most cases. Ways to combine data from various sources or that produced by the same source at different times is sometimes an important issue. This chapter describes the techniques that are most generally useful for the above purposes.

INTRODUCTION

This chapter discusses techniques useful for looking at data as a body of information. Ways will be discussed to decide whether all individuals in a set of data should be considered as members of a population, and hence can be combined to describe some aspect(s) of the population and/or used to make statistical decisions on various questions about the population. In this discussion, data points and the results of measurements produced by a group of laboratories will be considered as individuals in a population of laboratory outputs. This subject will be subdivided into "outliers" and "combining data sets" The treatment of interlaborarory collaborative test data also will be considered.

OUTLIERS

The question of "outliers" is one that often plagues both the producers and users of data. Given a set of data consisting of a group of numbers that all ideally should have the same value, how different do individual values have to be before they may be considered to be aliens, and not just extreme deviations from the mean?

As alluded to above, one is suspicious of individuals when, on ranking from the smallest to the largest value, either the smallest or the largest (sometimes data on each end) appear to be very much removed from their nearest neighbors. Another suspicious situation is when data are plotted and some point or points seem to deviate abnormally from the graphed line. What can be done to confirm or allay such suspicions?

Why is there so much concern about outliers anyway? There are two good statistical reasons. When outliers are included with legitimate data points, the calculated mean could be in error as a result. This is especially true in the case of small sets of data. Unfortunately, unless the data characteristics are well known from other sources of information, statistical rejection becomes less satisfactory in the case of small data sets. Second, outliers can greatly influence the calculated value for the standard deviation, again in the case of small data sets, and especially if the range is used for the estimation of the standard deviation.

There is another reason that is of interest to people that make measurements. Outliers indicate either mistakes, blunders, or malfunctions of the measurement process, and may indicate that corrective actions need to be taken to improve the system. In present day terminology, such behavior may be called "out of control" [1].

Whenever an outlier is suspected, the first prudent action to take is to critically look at all aspects of the measurement process that produced it. Of course, an arithmetic miscalculation should be the among the first things to be considered and checked. Then, the possibility of transcription errors in reporting data and even transposition of integers in the value reported need to be investigated. Malfunction of the measurement system is another possibility. Then there could just be an unexplainable mistake of some kind. Do not rule out the possibility that the wrong sample was measured!

Failing to find some reason for rejection, one of the procedures to be described below may be used to make a statistically supported decision with respect to rejection or retention of a suspected outlier.

The Rule of the Huge Error

If one has a reasonable idea (even an informed guess) as to what the standard deviation might be, then if the suspected point deviates from the mean by some predetermined multiple of it, it may be considered to be an outlier. In other words.

$$\frac{|\text{suspect} - \text{mean}|}{\text{"standard deviation"}} = M$$

If $M > 4$, then the suspect may be considered to be an outlier. The huge error rule is simply a crude t test. It is not clear what is the probability for such a decision, however. If the standard deviation assumed is well established, then the probability level could be as large as 999 in 1000 or more. If not, but it is a "ball-park" figure, the probability could be much closer to 19 in 20.

The huge error is applicable (though rarely used) for screening a data set for outliers, in which application the procedure to be used is obvious. One excludes the suspect and calculates a value for s. The difference of the suspect from the mean is compared with s to calculate a value for M. The decision to reject depends on a value of M > 4.

The huge error approach is especially useful in deciding whether a suspected point should be plotted along with other points in a data set. The data, neglecting the questioned point for the time being, are plotted and a "best fitting" line is drawn graphically, or better by a least-squares fit (see Chapter 7 for a discussion of curve fitting). The standard deviation of the fitted points or the average residuals from the fitted line are calculated and compared with the residual for the suspect point. If M > 4, reject the point. Otherwise, include it in the set and refit the line. No practical problems result from substituting the average deviation of all points from the line (excluding the suspected outlier) for the standard deviation when calculating an M value.

Example A set of data was taken to calibrate a measuring instrument. In the course of a prelimimary manual plot of the data (eye estimate), it was noticed that one point appeared to be much farther from the line than the others, so it was ignored for the moment. The analyst was tempted to reject it because he feared (and rightly so) that its inclusion (if it were an outlier) would falsify his least squares fit. The average deviation of the plotted points from the line (observed – curve) was 0.012. The deviation of the suspected point from the line was 0.06. Using the Huge Error Rule

$$M = \frac{0.06}{.012} = 5$$

The analyst decided to reject the data point.

The Dixon Test

The Dixon test [2,3] for outlying points in a data set is easy to use because of the simple calculations required. In using it, it is assumed that the population mean and standard deviation are unknown and that the data set on hand is the only source of information. A further assumption is that the data are normally distributed. The data points are ranked, serially, from the smallest to the largest. When this is done, a point at either end may appear to be considerably removed from its nearest neighbor, so that it is suspected of being an outlier. The Dixon test is based on the probability that such a situation could happen by chance, due to the variability of the data.

Use the following procedure:

1.   Rank the data in order of increasing size

$$X_1 < X_2 < X_3 < \ldots\ldots < X_n$$

2. Choose the confidence level for which rejection is merited.
3. Calculate the critical Dixon ratio, depending on the number of data points in the set.

Number of Points	Ratio to be Calculated
n = 3 to 7	r_{10}
n = 8 to 10	r_{11}
n = 11 to 13	r_{21}
n = 14 to 24	r_{22}

where the ratios are as follows:

R	If X_n Is Suspect	If X_1 Is Suspect
r_{10}	$(X_n - X_{n-1})/(X_n - X_1)$	$(X_2 - X_1)/(X_n - X_1)$
r_{11}	$(X_n - X_{n-1})/(X_n - X_2)$	$(X_2 - X_1)/(X_{n-1} - X_1)$
r_{21}	$(X_n - X_{n-2})/(X_n - X_2)$	$(X_3 - X_1)/(X_{n-1} - X_1)$
r_{22}	$(X_n - X_{n-2})/(X_n - X_3)$	$(X_3 - X_1)/(X_{n-2} - X_1)$

4. Look up the critical value for the appropriate r in the table (e.g., Table A.7 in Appendix A), for the confidence level of the test
5. If the calculated value is larger than the critical value in the table, reject the suspect point, otherwise retain it.

Example   Consider the following example to illustrate the Dixon test. Given the already ranked data set

$$9, 12, 12, 13, 13, 14, 15$$

The value 9 is suspect.
  Since there are 7 points, r_{10} is calculated

$$r_{10} = \frac{3}{6} = 0.500$$

Use the 95% confidence level for rejection. From Table A.7 in Appendix A, the critical value for $r_{10} = 0.507$. The decision is to retain the data point.
  Look at the table and note that, for a 90% level of confidence, the critical value of r is 0.434. From this one might conclude that rejection could be made on the basis

of more than a 90% confidence level but less than a 95% confidence level – say approximately a 94% confidence level.

Suppose that X_1 had a value of 8. One would then have calculated

$$r_{10} = 4/7 = 0.571$$

This is larger than the critical value of 0.507 and rejection could be done with 95% confidence. Remember that (1 – confidence) is the risk of false rejection. This means that though there is a 95% confidence that rejecting was merited, there is also a 5% risk that the wrong decision was made. Chapter 9 gives more information concerning the possible errors in hypothesis testing.

If a point is rejected one should look at the remaining data to decide whether a second point is suspect. If so, repeat the procedure. In a good data set, the chance for a second rejection should be small, and a third rejection even smaller. If multiple rejections are indicated, the measurement process is probably not in control [1].

While the Dixon test is easy to use, one should not get overly zealous about using it or any other outlier test, for that matter. Data represent work and money and neither should be discarded lightly. The analyst should try to salvage data if at all possible. Also, the presence of outliers should be a warning that perhaps the system has problems that should be solved by other procedures rather than by cavilierly discarding data.

The Grubbs Test

Another test for outliers that is widely used is the Grubbs test [3,4]. This test requires the calculation of the sample standard deviation and is thus a little more laborious than the Dixon test. However, even simple modern pocket computers can do this computation easily, so this is not a significant objection for its use. The procedure to be used is as follows:

Step 1.   Rank the data points in order of increasing value.

$$X_1 < X_2 < X_3 < \ldots \ldots < X_n$$

Step 2.   Decide whether X_1 or X_n is suspect

Step 3.   Calculate the sample average and standard deviation using all of the data for the moment, i.e. obtain $\overline{X}$ and s

Step 4.   Calculate T as follows

$$\text{If } X_1 \text{ is suspect}, T = (\overline{X} - X_1)/s$$
$$\text{If } X_n \text{ is suspect}, T = (X_n - \overline{X})/s$$

Step 5.   Choose the level of confidence for the test and compare the calculated value of T with the critical value in the table (such as Table A.8 of Appendix A).

If the calculated value exceeds the critical value, reject, otherwise retain the data.
Example Consider the same data set used for the Dixon Test, namely,

$$9, 12, 12, 13, 13, 14, 15$$

Again, 9 is the suspected point.
    Calculation of the standard deviation gives

$$\overline{X} = 12.57 \quad s = 1.90$$
$$T = 3.57 / 1.90 = 1.87$$

For 95% confidence of rejection, $T = 1.938$ from Table A.8 in Appendix A. The
decision is to retain the data point. If X_1 had been 8, as was postulated above, then

$$\overline{X} = 12.43 \quad s = 2.22$$
$$T = 4.43 / 2.22 = 1.99$$

In this case, one could reject at the 95% confidence level.

Youden Test for Outlying Laboratories

    Every laboratory that produces data should be concerned as to how its perform-
ance compares to that of other good laboratories. When the data of various laboratories must be used, interchangeably, that concern is shared by the data users, as well.
How can peer performance be judged with respect to bias and precision? Youden [5]
addressed the first question a number of years ago, which is the subject of the present
section. The question of precision was addressed by Cochran [6], and that will be
discussed in the next section.
    The Youden test consists in judging laboratory performance on the basis of the
results of several round-robin tests. The laboratories in question are given the same
material and asked to measure some property a prescribed number of times (even a
single measurement). The laboratories are each given a number of materials at one
time (which measures short-term performance) or the round robin is conducted a
number of times (which measures long term performance). For each material, the
laboratory reporting the highest test result is given the score of 1, the next highest the
score of 2, and so forth. The scores for each laboratory are summed and compared
with tabulated values, based on probability considerations. All laboratories *must*
make the same number of measurements.
    Obviously, if a laboratory always produces the highest or the lowest result, there
would be no doubt that it is a consistent producer of such. However, what about a
laboratory that produces such results most of the time? Where is the cut-off where
chance takes over? Youden's table (Table A.9 in Appendix A) gives the range of

scores that would be expected by chance alone with 95% confidence (i.e., 95% of the time). Scores outside of that range have only a 5% probability of being due to 'the way the ball bounces'. The range depends on the number of laboratories and the number of materials that were used in computing the cumulative scores.

Suppose that 10 laboratories participated and each analyzed the same 10 test materials. The results are scored and the cumulative scores for each laboratory are calculated as already described. Referring to Youden's Table A.9 in Appendix A, there is a 95% probability that the scores will lie within the range of 30 to 80, due to chance alone. There is only a 5% probability that scores lower than 30 would be due to chance, hence this would indicate higher than group performance. Scores in excess of 80 likewise would indicate lower than group performance.

Just as in the case of other statistical tests, small data sets are difficult to judge by probabilistic considerations. Notice the blank spaces where the test cannot be used. Also, there are spots where all test results must be high or low before judgment of outliers can be made. Judgment is best made on the basis of large amounts of data, i.e., a large number of laboratories measuring a large number of materials. However, this may not be feasible in many cases.

While an outlying data point ordinarily means an error or some problem, an outlying laboratory merely means performance that differs from the group. In the experience of the author, there have been a number of times when the outlying laboratory was the one producing correct data while the others were wrong. Reasons for outlying performance of laboratories thus need to be investigated, just as in the case of individual data points. The only way to tell whether accurate results are being obtained is to compare reported values with known values for the test materials. This has been discussed earlier.

Example

Laboratory	Sample Results				
	1	2	3	4	5
L	11.6	15.3	21.1	19.2	13.4
M	11.0	14.8	20.8	19.3	12.8
N	11.3	15.2	21.0	18.9	12.8
O	10.8	15.0	20.6	19.0	13.3
P	11.5	15.1	20.8	18.6	12.7
Q	11.1	14.7	20.5	18.7	13.0
R	11.2	14.9	20.7	18.8	13.2

For 7 laboratories analyzing 5 materials, the range of cumulative scores is expected to be within 8 to 32. Accordingly, Laboratory L is considered to consistently produce high results, when compared with the other six laboratories.

	Sample Scores					
Laboratory	1	2	3	4	5	Σ
L	1	1	1	2	1	6
M	6	6	3	1	6	22
N	3	2	2	4	5	16
O	7	4	6	3	2	22
P	2	3	4	7	7	23
Q	5	7	7	6	4	29
R	4	5	5	5	3	22

Cochran Test for Extreme Values of Variance

The F test already described is useful when deciding whether two variances are significantly different. The Cochran test [6] is used to decide whether there is an extreme (large) variance in a group of variances reported by the same laboratory on a number of occasions, or by a group of laboratories. The latter situation is where the Cochran test is used most frequently, such as in decisions concerning round-robin tests, for example. The only restriction for the test is that each variance considered must be based on the same number of degrees of freedom.

The procedure to be used is the following:

Step 1.   Calculate each variance and rank the values obtained from smallest to largest. Only the largest estimated variance is of concern.

Step 2.   Calculate the ratio

$$s^2(\text{largest})/\sum s_i^2$$

Step 3.   Compare the ratio with the value tabulated in a table such as Table A.10 in Appendix A. If the calculated ratio exceeds the tabulated value, consider the largest variance an extreme value with 95% confidence (5% risk of wrong decision).

It will be noted that the critical value depends on the number of variances compared and the number of replicate values used to estimate each variance. Remember that each variance must be based on the same number of measurements, in order to use this test.

Example   A group of laboratories in a round robin reported the following results, based on triplicate measurements:

Laboratory	1	2	3	4	5
s	1.25	1.33	1.05	2.75	1.10
s^2	1.56	1.77	1.10	7.56	1.21

Cochran statistic = 7.56/13.20 = 0.5727

The critical value from Table A.10 in Appendix A is 0.6838. Conclude that the value found by laboratory 4 is not an outlier. If laboratory 4 had reported s = 4.0, it would have been considered an extreme value, while s = 3.0 would not have been so considered. Perhaps the reader might like to calculate what would have been the maximum value that would not have been considered to be extreme.

MINITAB Example

Although MINITAB does not compute values for the Dixon, Grubbs, Youden, or Cochran tests, it is possible to quickly identify potential outliers using boxplots. Although ranking data from small datasets is not difficult, the task becomes much more tedious for large datasets, making the computer necessary. One can then apply the previously discussed tests to determine whether an identified value should actually be considered an outlier. Refer back to the MINITAB acid titration data, 'acid.mtw' that was introduced in Chapter 5. To assess the possibility of outliers in the data, it is helpful to examine boxplots of the titration data for each of the two acids separately. This is performed by the following menu commands:

Under Graph, Boxplot
- Under Multiple Y's, select **Simple**, click OK
- in the **Graph Variables** field
     double click to add C1 Acid 1 and C2 Acid 2 under **Graph Variables**
- under **Labels,** add a title in the **Title** field, click OK, click OK

This provides the boxplots of the acid titration data shown in Figure 6.1.
The asterisks denote possible outliers in the data. These are defined as values lower than (Q1-1.5{Q3-Q1}) and above (Q3+1.5{Q3-Q1}) where Q1=25th percentile and Q3=75th percentile. From the graph it is apparent that the smallest 6 and largest 6 observations from acid 1 and the smallest 2 observations from acid 2 are suspect. The exact values can be quickly identified by sorting the data using MINITAB menu commands:

Under Data, Sort
- double click on C1 Acid 1 to add it to the **Sort Columns** field
- click **Columns of Current Worksheet** and enter C5
- double click on C1 Acid 1 to add it to the **By Column** field, click OK

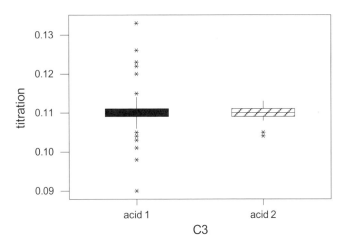

Figure 6.1. Boxplot of titration data.

Column 5 in the spreadsheet now has acid 1 titration data sorted in ascending order. This facilitates identification of the largest and smallest observations.

COMBINING DATA SETS

There are many times when one might want to combine data sets in order to compute a better estimate of a standard deviation or of a population mean. It has already been discussed how to do this in the case of the standard deviation, namely to compute a pooled standard deviation (see Chapter 4). This should be preceded by an F test (see Chapter 5) when pooling two s values, and by the Cochran test [6] when several values are concerned.

The present section will be devoted to guidance for the pooling of means. As in the case of s values, one should be confident that one is dealing with compatible or peer data. The rationale for dealing with such data is given in Figure 6.2.

The first question to be answered is whether confidence limits have been assigned or are assignable to each set of data that is to be combined. All data that do not meet this requirement should be rejected as 'quality unknown'.

Data for which confidence limits are available should be examined for significance of any apparent differences. They may be examined by one of the outlier tests already given or by the test given in Chapter 9. While an outlier may be rejected, the reason for its status as an outlier should be investigated and verified if possible. Remember, the outlier could be the correct value! All compatible data should be weighted according to one of the three procedures that will now be discussed. Note that the formulas given to carry out each procedure include one that may be used to calculate the standard deviation of the weighted mean in each case [7].

Case I: All Data Believed to Have the Same Precision

Case I A: Same Number of Measurements in Each Set

Compute a simple mean of the means, that is, a grand average or grand mean.

$$\overline{\overline{X}} = \frac{\overline{X}_1 + \overline{X}_2 + \ldots + \overline{X}_k}{k}$$

$$s_{\overline{\overline{X}}}^2 = s_{\overline{X}}^2 / k \text{ or } s_{\overline{\overline{X}}} = \sqrt{(s_{\overline{X}}^2 / k)}$$

Case I B: Same Precision but Unequal Number of Measurements in the Respective Means

Compute a weighted mean in which each mean is multiplied, i.e., weighted, by its corresponding number of measurements, W_i. These values are summed and divided by the sum of all of the weights (total number of measurements).

$$\overline{\overline{X}} = \frac{\overline{X}_1 W_1 + \overline{X}_2 W_2 + \ldots + \overline{X}_k W_k}{W_1 + W_2 + \ldots + W_k}$$

$$s_{\overline{X}}^2 = s_{\overline{X}}^2 / (W_1 + W_2 + \ldots + W_k)$$

$$s_{\overline{\overline{X}}} = \sqrt{[s_{\overline{X}}^2 / (W_1 + W_2 + \ldots + W_k)]}$$

Case II: Data Sets Have Different Precisions Due to Number of Measurements and Differing Variances

The values are weighted inversely according to their variances [7], i.e.,

$$W_i = 1/V_{\overline{X}i}$$

$$\text{or } W_i = n_{Xi} / s_{Xi}^2$$

$$\overline{\overline{X}} = \frac{\overline{X}_1 W_1 + \overline{X}_2 W_2 + \ldots + \overline{X}_k W_k}{W_1 + W_2 + \ldots + W_k}$$

$$s_{\overline{\overline{X}}} = \sqrt{[1/(W_1 + W_2 + \ldots + W_k)]}$$

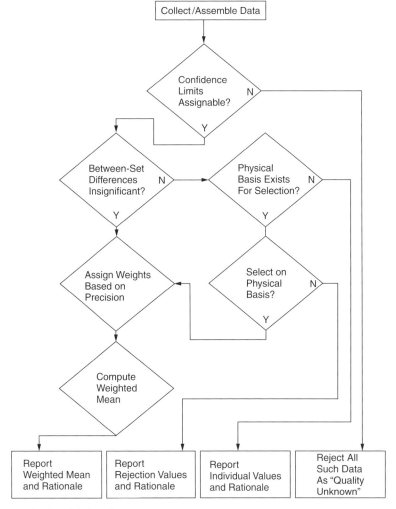

Figure 6.2. Combining data sets.

The reader is no doubt familiar with Case I, so no examples will be worked out.
Example Since Case II may be less familiar, the following example is included to
clarify the procedure.

Set no. (i)	$\overline{X}_i$	n_i	s_i	W_I
1	10.50	10	.10	1000
2	10.37	5	.15	222
3	10.49	20	.11	1652
4	10.45	5	.10	500
5	10.47	7	.16	273

$$\overline{X} = \frac{10.50\times1000+10.37\times222+10.49\times1652+10.45\times500+10.47\times273}{1000+222+1652+500+273}$$

$$\overline{\overline{X}} = 10.478$$

$$s_{\overline{X}}^2 = 0.000274 \qquad\qquad s_{\overline{X}} = 0.0166$$

STATISTICS OF INTERLABORATORY COLLABORATIVE TESTING

Collaborative testing consists of exercises in which a group of laboratories jointly participate for one of the following purposes:

- To validate a specific methodology
- To evaluate their respective proficiencies in performing a specific method of test
- To estimate a consensus value of some parameter of a reference material

While there may be other reasons for a collaborative test, the ones mentioned above are the ones most frequently encountered and that will be discussed in the following sections.

Validation of a Method of Test

Both standardization organizations and regulatory agencies conduct collaborative tests to validate methods. The procedures used vary in detail but essentially follow the procedure that will be described below. The reader should consult the referenced articles for details of the procedures used by the several organizations cited [8-10]. The philosophy of collaborative testing is discussed in a paper by the author [11] and in his book on quality assurance of chemical measurements [1].

The number of laboratories participating may vary, but a minimum of six is required in most cases. Ordinarily, the method is applied to at least three test levels of the analyte determined, unless the test is intended for a single level or limited concentration range. Each laboratory is instructed to measure each test sample a definite number of times. Duplicate measurements are often specified and the procedure outlined below is for such a case. Each laboratory must adhere to the measurement plan with respect to following the procedure and the number of replicate measurements made. A committee or other specified persons (or a referee) reviews and analyzes the test results.

Typically, the test results are screened initially for outlying results using the tests described earlier. The Cochran test is used to screen for extreme variance. The Dixon or Grubbs test is used to identify outlying test results. In either case, effort is made to salvage discrepant data by checking for such blunders as miscalculations or wrong reporting units, for example.

The data are statistically analyzed for precision, using the procedure described in Chapter 4, which the reader should consult with respect to the following discussion.

The within-laboratory standard deviation, s_w, is estimated from the expression (in the case of duplicate measurements)

$$s_w = \overline{R}/d_2^* \text{ or } \sqrt{\sum (d^2/2k)}$$

The between laboratory standard deviation, s_b, is estimated by the expression

$$s_b = \sqrt{(s_T^2 - s_W^2/n)}$$

where n (2 in the case of duplicates) is the number of replicate measurements made by each laboratory. The IUPAC recommends reporting the repeatability, r, of the test method as

$$r = 2\sqrt{2}\, s_w$$

and the reproducibility, R, as

$$R = 2\sqrt{2}\, s_b$$

These are said to represent the 95% confidence statistics for the agreement of two test results, made under the respective conditions. American standardization organizations do not follow this practice.

The evaluation of bias depends on the significance of the differences of the average test results from the known values for the test samples. If the average of all test results differs significantly from a known value, the method is said to produce biased results at that level of analyte measured.

Proficiency Testing

Proficiency testing ordinarily relates to tests of the performance of laboratories that must produce consistent results for regulatory or similar purposes. It is also used in qualifying laboratories for certification purposes.

In a typical exercise, all laboratories tested are given identical samples and their test results are compared with the known values for the test samples. The degree of deviation permitted may be assigned, arbitrarily, or it may be calculated on the basis of a pooled standard deviation of participating laboratories. In either case, what amounts to an engineering tolerance interval ($\pm z \sigma$) (see glossary in Appendix B) is set around the value (either a 95% or a 99% interval) and laboratories are judged to perform acceptably if their results fall within the interval.

Testing to Determine Consensus Values of Materials

A group of laboratories may be utilized to collaboratively assign consensus values to reference materials. Each participating laboratory analyzes the test material following a prescribed test plan. While the details of a suitable test plan are not within the scope of the present book, a few comments are in order to indicate the general practices that should be followed. Each laboratory should make the same number of measurements. Measurement sequences should be randomized to minimize bias due to order of measurement. Replications should be included to evaluate material homogeneity and laboratory precision as was described in Chapter 4. The plan should detail calibration procedures and include the measurement of already certified reference materials as possible to be used to evaluate any interlaboratory biases that might occur.

The test results are evaluated according to the procedure described in the section Combining Data Sets. Outliers should be eliminated when identified by the tests described earlier in this chapter. The average value and the standard deviation for the results of the test should be calculated as described. A confidence interval for the test result can be calculated as described earlier, using the degrees of freedom coresponding to the number of determinations.

RANDOM NUMBERS

Random numbers are very useful in measurement programs to minimize the chance for introduction of uncertainties, due to systematic treatment of the data. One may want to randomize the selection of samples to be measured or to randomize the order in which certain measurements are made or certain operations are carried out. One cannot think randomly, no matter how hard one might try. In fact, the harder one tries, the more unlikely one is to think randomly. A random process must be used to accomplish the various purposes mentioned above.

Random numbers can be generated several ways. One could draw cards from a well-shuffled deck or cast dice. The familiar 6-sided die may be the reader's first thought, but multifaceted dice, for example, 20-sided dice are also available [12]. Some hand-held calculators are capable of generating random numbers.

A table of random numbers, such as Table A.11 in Appendix A, is a time-honored way to obtain random numbers. While extensive tables are available, the table given here should be satisfactory for many purposes.

No matter what the procedure, the first step is to assign numbers to the objects, samples, areas to be sampled, or to the measurements to be made. The selection process is carried out according to the random numbers. This will be illustrated by the use of random number Table A.11 in Appendix A.

Enter the table by a random process. This could be by asking a colleague to give a number to be used for column selection and another for use for row selection, or the table itself could be used to do this selecting. For example, with closed eyes, move a pointer around on the page. The number it points to will be the column number. Repeat this operation to get a row number. If an unusable number is indicated, repeat until a useable one is found.

Once the starting place is found, proceed in a systematic direction from this position. The strategy to be used should be decided beforehand. The direction could be like reading a book, or reverse reading, or one could move down from column to column, or up if desired. Once this decision is made, the pattern is followed until the random number selection is complete.

Example It is desired to measure 10 samples of material according to a random sequence. The samples are first assigned serial numbers ranging from 01 to 10. Assume that column 5 and row 11 have been chosen as starting places by a random process, for entry to the table. Proceed from that point like reading a book. The number at point of entry is 16, which cannot be used, since it is not one of the samples. The next number is 07. Sample 07 will be the first sample analyzed. Passing successively through 68, 14, 97, 17, 57, 59, 93, and 81, all numbers that cannot be used, the next number is 08; hence sample 08 will be the second sample to be measured. Proceeding in the same way, the rest of the numbers may be found. If one comes to a number that has already been chosen, it is passed over to the next unchosen number.

In this example, the sequence found was

Sequence No.	1	2	3	4	5	6	7	8	9	10
Sample No.	07	08	02	04	10	01	05	03	09	06

If the problem had been to randomly select 10 samples from a group of 100, the procedure would be similar. Having randomly selected the starting place, one would move from it in a pre-selected systematic way and analyze samples as their numbers are encountered in the table.

One could randomly measure the randomly selected samples by choosing an order as outlined above.

MINITAB Example

One advantage of computer software is that the user has access to a virtually unlimited supply of random numbers. These may be used to determine the order of a procedure (as described above) or in a much more sophisticated technique called simulation. Simulation is discussed in more detail in Chaper 9. MINITAB allows the user the ability to generate random numbers from a variety of statistical distributions. Each distribution has its own distinguishing features that make it appropriate in certain situations. The Table 6.1 details several of the most common distributions, their features, and circumstances for use.

Table 6.1. Random Number Distributions

Distribution	Main Features	Possible Situations for Use
Normal	• Symmetric and bell-shaped distribution • Continuous data	Generate random numbers representing the grades of a class on an exam
Gamma	• Skewed, positive distribution • Continuous data	Generate random numbers representing the survival times of infected rodents
Bernouli	• Gives a success or failure outcome for 1 event • Discrete data	Generate a random number representing whether 1 coin toss yields a head
Binomial	• Gives the total number of success outcomes in n events • Discrete data	Generate a random number representing the total number of heads when a coin is flipped 10 times
Hyper-geometric	• Gives the number of successes in a sample from a population • Discrete data	You have an urn with 10 balls in it (8 black, 2 red). Randomly sample 4 balls. Generate a random number representing the number of red balls in the 4 ball sample
Integer	• Equal probability over the interval of definition • Discrete data	Generate a random number representing the outcome when 1 die is rolled
Poisson	• Gives the number of counts observed on each of n events • Discrete data	Generate a random number representing the number of words on a page in a book

Note that the above table is not meant to detail statistical properties of these distributions, but rather descriptive information for non-statisticians. Precise statistical definitions for the above distributions may be found in [13]. Properties of the distributions (called parameters) required to generate the random numbers must be entered by the user. As with most statistical techniques, the major hurdle for the user is to become familiar with the distributions available in order to select the one most appropriate for a given situation. Although not terribly important for ordering, more advanced statistical techniques such as simulation require correct distributional assumptions for accuracy.

Example A metrologist has 100 measurements of radiation at a reactor. He knows that 97% of the time, the measurement device yields an accurate reading. If he samples 15 of the 100 measurements, generate a random number representing the number of accurate readings in the 15 samples.

This problem is asking for the number of successes (accurate readings) from a 15 reading sample given that the population has 97 successes on 100 trials (readings). This data is thus suited to the hypergeometric distribution.

The MINITAB menu commands to generate a random number of successes from 15 trials based on a population of 100 trials with 97 successes are as follows:

Under Calc, Random Data, Hypergeometric
- enter 1 under **Generate Rows of Data** field
- enter C1 under **Store in Columns** field
- enter 100 under **Population Size** field, 97 under **Successes in Population** field, and 15 under **Sample Size** field, click OK

The result is a 15 entered in column 1 of the spreadsheet. This information is indicating that 15 of the 15 readings taken would be accurate. Given the 97% success rate in the population for this data, this is not surprising! The reader should investigate how lowering the 'successes in population' field changes the random number generated representing successes in the sample. Several random numbers may be generated with one MINITAB command by changing the Generate Rows of Data field.

Example A metrologist would like to randomly generate 100 observations representing the amount of an air toxin measured (in micrograms). She knows from previous information that the data follows a normal distribution with mean 10.295 micrograms and standard deviation 2.269 micrograms. The following MINITAB menu commands produce the observations:

Under Calc, Random Data, Normal
- enter 100 under **Generate Rows of Data** field
- enter C2 under **Store in Columns** field
- enter 10.295 under **Mean** field and 2.269 under **Standard Deviation** field, click OK

Column 2 in the MINITAB spreadsheet now contains the data representing the toxin measurements.

Example A clinical trial coordinator is involved in a study of 3 different drug doses to be administered to each of 25 patients. He would like to randomize which treatment a patient starts with and wants each of the 3 treatments to be assigned with equal probability.

This is an example of the Integer Distribution on the interval 1-3, which requires the probability of generating a 1 to be equal to that of generating a 2 or a 3. The 25 uniform random numbers (on the interval (1,3)) are produced using the following MINITAB menu commands:

Under Calc, Random Data, Integer
- enter 25 under **Generate Rows of Data** field
- enter C3 under **Store in Columns** field
- enter 1 under **Minimum Value** field and 3 under **Maximum Value** field, click OK

Column 3 in the MINITAB spreadsheet now contains the data representing the first treatment for each of the 25 patients.

EXERCISES

6-1. Why is the question of outliers an important consideration? Answer from the point of view of their influence on a data set, the danger of false rejection, and the problem of false retention.

6-2. Discuss the similarities and differences in the Dixon test and the Grubbs test for decisions on possible outlier data.

6-3. What is Youden's test for outlying laboratories? How does it function? What do the conclusions really mean?

6-4. Screen the following data sets for outliers using both Dixon's and Grubbs' tests:

a. 14.5, 14.9, 15.3, 14.7 14.8, 14.9, 14.6
b. 7.1, 8.3, 8.4, 8.0, 8.9, 8.2, 8.0, 8.5, 8.3, 8.7

Use MINITAB to produce boxplots of each dataset and identify outliers.

6-5. Five laboratories, participating in a collaborative test, reported the following results on triplicate measurements of cyanide in water (all results in mg/L).

Laboratory	1	2	3	4	5
$\overline{X}$	.905	.957	.900	.926	.917
s	.062	.049	.050	.053	.056

a. Are any of the values of $\overline{X}$ outliers (95% confidence)? Use two tests.

b. Are any of the values of s outliers (95% confidence)?
c. Compute the pooled standard deviation for the test.
d. Compute the grand average of the cyanide values for the test.
e. Knowing that the test material contained 0.95 mg/L CN, does the test give biased results (at 95% confidence level)?

6-6. Compute a grand average for the following data sets:

$\overline{X}$	N	s
10.35	10	1.20
12.00	5	2.00
11.10	7	1.50

6-7. Ten laboratories were each given the same five samples to analyze to determine whether all of the laboratories could produce equivalent results. The following data were reported:

Lab. No.	1	2	3	4	5
			Sample Number		
1	5.59	2.46	4.64	3.19	7.32
2	5.94	2.52	4.68	3.28	6.44
3	5.80	2.40	4.62	3.12	6.89
4	5.73	2.46	4.65	3.09	7.17
5	5.72	2.51	4.62	3.12	7.00
6	5.80	2.51	4.80	3.29	7.48
7	5.45	2.40	4.45	3.07	7.02
8	5.72	2.50	4.58	3.27	6.76
9	5.63	2.32	4.69	3.04	6.92
10	5.88	2.42	4.67	3.16	7.39

Use the Youden test to decide whether all of the laboratories are peers, or whether any laboratories consistently produce high or low results with respect to the others in the group.

6-8. A batch of 50 samples was submitted for analysis. It was mutually agreed that the analysis would be made on ten randomly selected samples. Use Table A.11 in Appendix A for this purpose, devise the sample selection scheme, and provide enough information that the basis for your selection may be verified.

6-9. As a further precaution to minimize bias due to order of measurement, the samples in 6-8 should be measured in a random sequence. Devise such a sequence, using Table A.11 in Appendix A. Provide enough information so that your basis for selection may be verified. Devise a random sequence using random numbers generated in MINITAB.

6-10. Generate 50 random numbers representing the integer grades of high school students on a final exam. The grades should be out of 100. Use computer software of your choice.

6-11. Using the MINITAB dataset choleste.mtw, generate boxplots of the variable 2-Day to identify outliers. Identify outliers using the Rule of the Huge Error and the Dixon Test.

REFERENCES

[1] Taylor J.K. *Quality Assurance of Chemical Measurements* (Chelsea, MI: Lewis Publishing Co., 1987).

[2] Dixon W.J. "Processing Data Outliers," *Biometrics* 9(1): 74-89 (1953).

[3] "Standard Recommended Method for Dealing with Outlying Observations," (Philadelphia, PA: ASTM E-178, ASTM).

[4] Grubbs, F. E., and G. Beck. "Extension of Sample Sizes and Percentage Points for Significance Tests of Outlying Observations," *Technometrics* 14(4): 847–854 (1972).

[5] Youden, W.J. "Measurement Agreement Comparisons," in *Proceedings of the 1962 Standards Laboratory Conference,* NBS Miscellaneous Publication 248; (Washington, D.C.: National Bureau of Standards, 1962).

[6]   Cochran, W.G. "The Distribution of the Largest of a Set of Estimated Variances as a Fraction of the Total," *Ann. Eugenics,* 11: 47-53 (1941).

[7]   Ku, H.H. "Statistical Concepts in Metrology," in *Precision Measurement and Calibration: Statistical Concepts and Procedures,* H.H. Ku, Ed., NBS Special Publication 300, Vol. 1, (Gaithersburg, MD: National Institute of Standards and Technology, 1969).

[8]   Association of Official Analytical Chemists, Suite 400, 2200 Wilson Boulevard, Arlington, VA 22201-3301. (Note: The January 1, 1988 issue of the JAOAC contains a complete prescription for conducting a collaborative study.)

[9]   "Standard Practice for Determination of the Precision and Bias of Methods of Committee D 19 on Water," ASTM D-2777, (Philadelphia, PA: ASTM).

[10]   "Practice for Conducting an Interlaboratory Test Program to Determine the Precision of Test Methods," E-691, (Philadelphia, PA: ASTM).

[11]   Taylor, J.K. "The Role of Collaborative and Cooperative Studies in Evaluation of Analytical Methods," *J. Assoc. Off. Anal. Chem.,* 69: 393 (1986).

[12]   Lansford Publishing Co., P. O. Box 8711, San Jose, CA 95155; also; Technovat, 910 Southeast 12th Avenue, Pompano Beach, FL 33060.

[13]   Devore, J.L. *Probability and Statistics for Engineering and the Sciences* (Monterey: Brooks/Cole Publishing Company, 1987).

CHAPTER 7

Presenting Data

Data are commonly presented in tables, charts, and graphs and by mathematical expressions. The latter may be theoretical or empirical. This chapter consists of a brief discussion of these modes of presentation. Some of these modes provide ways not only to facilitate ready access to data but also to smooth out experimental variability in finished data. Excellent discussions of data presentation will be found in the literature, especially that of a number of years ago. Modern authors have dealt less extensively with this subject. The author grew up with the books cited in the following references and recommends them for their practical approach and ease of understanding of the subject matter presented [1,2].

TABLES

The use of tables is perhaps the most common method for presentation of data. The format will vary, depending on what information is needed to be conveyed. Even a cursory perusal of the scientific literature will reveal many examples of both good and poor tables. A good table is simply one that presents data in an easily understandable manner. Tables should be relatively simple in order to promote understanding and the columns should have a clear relationship to each other. Column titles should be as brief as possible, consistent with clarity. Footnotes may be needed in some cases to provide further explanation of the headings.

Columns that are intended to be intercompared should be placed adjacently as possible. Unnecessary grouping of dissimilar materials in tables can lead to confusion, hence this should be avoided whenever possible. For example, the author has seen examples where several kinds of uncertainty estimates have been included in the same table for different substances, for apparently no reason. When the statistical parameters of the results for all substances in a given table are treated identically, intercomparisons are facilitated. Moreover, there is an advantage in achieving consistency among all tables in a single presentation, such as a paper or a lecture, and this should be done unless there is good reason to do otherwise. In the latter case, pointing out the reason for the varied presentations will give emphasis to the reasons therefore, and help to minimize confusion of intercomparison.

In tables consisting of values of a homogeneous nature, for example the boiling point of a compound as a function of pressure, the intervals between numerical values should be sufficiently narrow to permit simple interpolation, whenever this is possible.

CHARTS

The use of charts is a time honored way to present compilations of data. The two most familiar ways are described in the following sections.

Pie Charts

A pie chart, consisting of sectors of a circle, is a popular means of presentation when it is desired to show the relation of several factors as fractions or percentages of the whole. There are many software packages that can display and print such presentations so that their use is now commonplace. Exploded pie charts in which one or more sectors is slightly withdrawn to emphasize relationships can be created very easily by modern computers. Although pie charts are not used frequently for presentation of scientific and technical information, a few words concerning their general nature may be in order. A typical pie chart is shown in Figure 7.1.

Pie charts are more comprehensible as the sectors are approximately equal. A feeling of relationship is lost as very small sectors are placed alongside very large ones. In any case, numerical values need to be inserted in the sectors or related to them by lines or arrows to provide numerical significance, since the eye is not a good quantitative judge of the relative areas of sectors.

The total number of sectors used should be reasonably small. While not a hard and fast rule, a maximum of eight sectors is a reasonable number. Sectors may be homogeneous or consist of conglomerates of several items. The information contained in a sector may be displayed as a separate pie chart. This is an effective way to handle conglomerates.

Bar Charts

Bar charts are useful when differences of several quantities are desired to be presented. The eye can see even small differences in relative heights so this is an effective medium to present such information. Data varying with time or location are favorite subjects for bar charts. Several bars can be displayed next to each other in several groupings to illustrate how several functions change with time and with respect to each other, for example. Such a chart is given in Figure 7.2.

The modern computer is again a master displayer of bar charts. LOTUS 1-2-3 [3], for example, can prepare bar charts as well as pie charts almost instantaneously, once the data have been entered into an appropriate spread sheet. A few key strokes is all

Elemental Gasoline Analysis

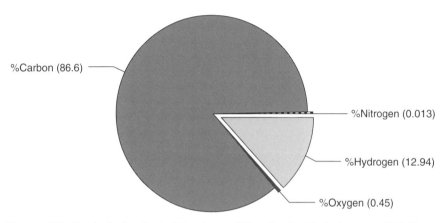

Figure 7.1. Typical pie chart. (Courtesy of Dr. G. Ouchi, Laboratory PC Users, San Jose, CA.)

that is needed to alter a format and new charts can be prepared automatically when data compilations are revised or when new data are entered into data bases.

Frequency distributions, commonly called histograms, are special kinds of bar charts that are used widely for displaying variability of scientific and technical information. Such displays may be used to demonstrate that a normal distribution is or is not achieved, for example. See Figure 3.7 in Chapter 3 for examples of such charts. Generally, a minimum of 25 data points is required to prepare a good bar chart, and considerably more is highly desirable. The data are divided into groups bounded by cells of fixed limits. The number of cells chosen to cover the range of values for the data is somewhat arbitrary. If too few, a distribution can lack resolution; if too many, there can be numerous unpopulated cells in the case of small data sets. Trial and error may be used in a specific case to decide what is most effective. Of course, the width of all cells must be the same in a given presentation. The ASTM Manual on Presentation of Data and Control Chart Analysis [4] suggests that somewhere between 10 and 20 cells is the proper number.

Once the number of cells and the cell boundaries have been established, the number of data points lying within each cell are tabulated and the heights of the cells are drawn accordingly. The cells may be situated contiguously (often the case) or represented as narrow bars or even lines, located at the midpoints of the cells. A normal distribution envelope may be superimposed upon the presentation for comparison purposes. See Chapter 9 for instructions for constructing this envelope. It

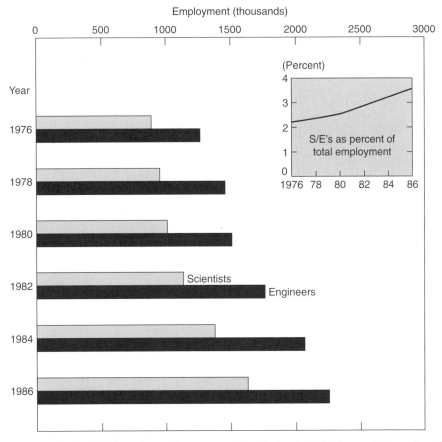

Figure 7.2. Typical bar chart. (Courtesy of Dr. Richard C. Atkinson, University of California, San Jose.)

should be noted that a normal distribution envelope can be superimposed on any histogram. How well it fits the histogram is always a matter of judgment.

GRAPHS

Graphical presentation is one of the most frequently used means to present scientific and technical data. The presentation may be for the purpose of showing relationships or trends, or for presentation of numerical values in a concise manner. Such a presentation can provide a means for interpolation of intermediate values, provided that the scale and rulings of the graph have sufficient resolution.

Graphs may consist of a single set or of several sets of data, each represented by its distinctive line or curve. Only related items should be included on the same graph and the number of lines should be restricted to promote clarity and ease of use. Scale factors should be chosen to facilitate interpolation and resolution of both axes. A slope of 45° of at least part of the data is desirable, unless a zero slope is normal. Many computer programs that draw graphs make such a selection automatically. When several graphs are drawn, a slightly different adjustment could be made for each, resulting in scales that are not equivalent. This should be kept in mind when intercomparing graphs. Suitable symbols or line characterization should be used when several data sets are displayed on the same graph.

A combination of graphical and tabular presentations may be used to good advantage. The former illustrates most effectively qualitative characteristics (e.g., changes of data with time or sequence) while the latter is the best means to present quantitative information.

Points plotted on a graph may be interconnected or represented by a smooth curve. The former approach should be restricted to nonfunctional relationships. Thus, variability and trends with time can be so shown. In fact, it would be misleading and hence improper to draw smooth lines to represent data that are not believed to be functionally related.

Points that are, or may be expected to be, functionally related should be represented by a line or curve rather than lines drawn through the plotted points. Not only does this practice present a best representation but it can effectively smooth measurement variability and make the data more useful. As has been mentioned earlier and will be pointed out later, a preliminary graphical representation of data can be highly recommended to identify outliers that would confuse numerical analysis. Some suggestions for plotting data are given in later sections. An excellent discussion of the use of computers in graphing data is contained in a recent book by Cleveland[5].

Linear Graphs

Whenever possible, plotted data should be represented by straight lines, drawn to best fit it. Straight lines are not only the easiest to fit but provide ease of use as contrasted to curved line representations. Because of measurement variability, data cannot be expected to fit a straight line perfectly, and chance errors can give the appearance of a curved distribution, even for data known to be linear. This is especially true when only a few points are available for use in establishing a line. Of course, if many well-distributed data points are available and a curvilinear pattern is evident, it would be improper to represent it in any other way.

An old, established practice, used less frequently today, was to bracket plotted points with error bars, corresponding to calculated or estimated limits of uncertainty. When included, a line drawn close to the points and through the error bars would appear to be a good fit of the data. In such a case, the deviations from the drawn line

Table 7.1. Some Linearizing Transformations

Transformed Variables Plotted to Give a Linear Relationship		Equation of the line	Intercept a′	Slope b′
Y_T	X_T		Corresponds to	
Y	1/X	$Y = a + b/X$	a	b
1/Y	X	$Y = 1/(a + bX)$	a	b
X/Y	X	$Y = X/(a + bX)$	a	b
log Y	X	$Y = ab^x$	log a	log b
log Y	log X	$Y = aX^b$	log a	b
Y	X^n	$Y = a + bX^n$	a	b

Fit $Y_T = a' + b'X_T$

Taken from M.G. Natrella, "Experimental Statistics" [9].

should be no larger and perhaps smaller than the variability of the individual points. In fact, the F test for goodness of fit, described later, is a statistical utilization of the principle described above.

Nonlinear Graphs

Data that clearly demonstrate a curvilinear distribution must be so represented on a graph. However, there are some alternatives that can be investigated. Nonlinear data may be linearized in some cases by plotting on non-linear coordinate paper (semi-log or log-log plots, for example) or by transforming the data before plotting. Some linearizing methods are presented in Table 7.1. Either the X values, the Y values, or both are transformed as indicated in the table. If a linear fit of the transformed data is applicable (by graphical fit), a function of the type

$$Y_T = a' + b'X_T$$

can be fitted. The table indicates how a′ and b′ are related to the constants a and b of the equation for the original data.

Another approach is to break up the data into two or more groups, each of which may be reasonably represented by a straight line.

In drawing a curved line to represent a number of points, it should be remembered that the most simple curve is again the best approach when possible. Changes in curvature ordinarily indicate some change in physical relations. An empirically fitted curved line may show changes in curvature when there really is no such change, rather the curvature is due to variability of the data.

Nomographs

Nomographs are effective ways to graphically calculate various functionally related quantities. Nomographs are really graphical computational devices. They were once used widely in engineering situations when calculating was more laborious than at the present time, and they still can be useful when complex relationships are concerned. In brief, scales are laid out in which the scale intervals and placement of the lines are chosen by well-established procedures. A straight edge can then be used to interconnect independent variables so the corresponding values of dependent variables can be read.

Recent examples of effective nomograms are contained in a paper by Provost [6]. In it, he presents nomographs to permit estimation of the number of samples and analytical measurements to be made, based on precision and cost considerations, to limit measurement uncertainty to an acceptable level.

One of the classical books that describe the construction of nomographs was written by Lipka [1] in 1918. Despite its early date, it remains a very important source of information on this subject. The examples cited by its author are almost endless and graphs reproduced are a delight to behold. The book may be difficult to find but the search will be worth the effort if one wants detailed information on preparing nomographs.

MINITAB Example

MINITAB offers users a wide variety of graphical capabilities with minimal effort. The graphical output can be easily copied and inserted into word processing software. The MINITAB dataset 'defects.mtw' details the number of defects found at various manufacturing plants, with information on the type of defect and the date. The owner of these plants wants to know what kinds of defects are most common. The pie chart summarizes this type of information nicely and is appropriate for depicting outcomes for one variable. The following MINITAB menu commands produce Figure 7.3, which shows each type of defect as a proportion of all defects.

Under Graph, Pie Chart,
- select **Chart Values from a Table**, double click on C3 Describe to add it to the **Categorical Variable** field

- double click C2 Defects to add it to the **Summary Variables** field
- under **Labels**, Slice Labels, select Percent, click OK, click OK

Note that there are many options for features such as adding titles, selecting fill types and colors of the pie, and enlarging a specific pie. The pie chart produced indicates that o-ring defects are most common and comprise 26% of the total defects.

Pie Chart of Manufacturing Defects

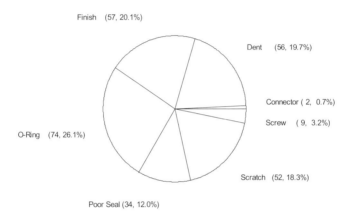

Figure 7.3. Pie chart of manufacturing defects.

Pie graphs are useful for depicting one type of information, such as type of defect. Other graphs, such as bar graphs and line graphs, can summarize more than one type of information. The MINITAB dataset 'cities.mtw' provides monthly temperature information for several cities in the United States. The data for New York and San Diego are in Table 7.2.

The menu commands

Under Graph, Scatterplot
- select **With Connect and Groups**, click OK
- in the **Graph Variables** field
    double click to add C4 New York and C1 Month under **Y** and **X** for Graph 1
    double click to add C5 San Diego and C1 Month under **Y** and **X** for Graph 2
- under **Multiple Graphs,** Multiple Variables, select Overlaid on the Same Page, click OK

- under **Labels**, add a title in the **Title** field, click OK, click OK.

produce Figure 7.4.

Table 7.2.   **Cities Dataset from MINITAB**

Month	New York	San Diego
1	32	56
2	33	60
3	41	58
4	52	62
5	62	63
6	72	68
7	77	69
8	75	71
9	68	69
10	58	67
11	47	61
12	35	58

Taken from MINITAB

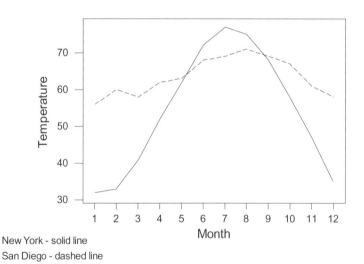

Figure 7.4. Linear graph of cities data.

Although adequate, it is more informative for the scale of months to show the actual month name. In earlier versions of MINITAB, this was achieved using menu commands before creating the graph. Version 14 allows the user to edit items such as axis labels and scales (among other features) after the graph is created by double clicking on the appropriate area of the graph. The modified monthly temperature graph is shown in Figure 7.5.

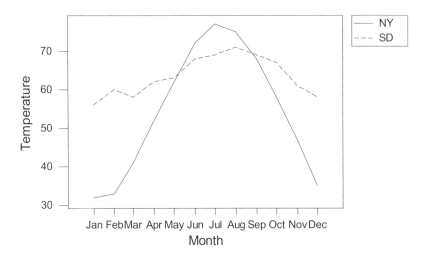

Figure 7.5. Linear graph of cities data-revised

MATHEMATICAL EXPRESSIONS

The fitting of a mathematical expression to data is desirable for several reasons. As will be seen later, the fit can be made quite objectively. A mathematical equation of appropriate form permits the numerical expression of results much more precisely than any other means. Furthermore, values of a dependent variable can be calculated for essentially any desired value of the independent variable. The labor of fitting has been reduced to such an extent by modern methods of calculation that mathematical fitting has become the method of choice.

Theoretical Relationships

In some cases, a functional relationship is known, based on physical or chemical theory. If so, the fitting is simplified in that the equation to be used is predetermined. The methods available for fitting are summarized in later sections.

Empirical Relationships

When theoretical relationships are not known, empirical relationships may be fitted to data with the advantages already mentioned. Some considerations to be made in choosing an appropriate empirical relationship are discussed below.

Linear Empirical Relationships

Linear relationships are of the form

$$\boxed{Y = a + bX}$$

For simplicity, they are to be preferred whenever it is possible to use them. The reasons are about the same as mentioned when graphical fitting was discussed. Experimenters long have emphasized that the most simple relationship is the one that should be used, and that most often is a straight line. The problem of selection of the function for use is complicated as the number of data points decreases (actually the degrees of freedom, namely the number of points minus the number of constants of the equation to be determined). Data with considerable variability also complicate decisions in that apparent curvature can arise due to chance variations.

A preliminary plot often will be helpful when deciding what mathematical relationship to use. If a linear fit does not seem appropriate, it is possible that transforming the data by one of the ways suggested in Table 7.1 will linearize it. Several modern computer programs will explore linearization approaches, automatically. For example, NWA STATPAK [6] and NWA QUALITY ANALYST [7] will fit a data set with linear, exponential, logarithmic, and power law expressions and draw residual plots to permit the user to decide which if any provides the best fit. MINITAB also offers these features.

No matter how a function is chosen, a residual plot provides an excellent means to evaluate the quality of a fit. The plot, of course, consists of a plot of the residuals with respect to the observed values. The residual referred to is

residual = observed − calculated from a fitted curve

If the residual is positive, the observed point is larger than the calculated point and vice versa. The residuals should be randomly distributed in the case of an acceptable fit. One should look for patterns (systematic departures) in residual plots which would indicate a poor fit of the experimental data. Calculation of residuals using MINITAB software will be discussed in an example at the end of the chapter.

Nonlinear Empirical Relationships

Any data may be fitted with a power series (generalized parabola) of the proper order. By this is meant an expression of the form

$$Y = a + bX + cX^2 + dX^3 + ...$$

Again experienced data analysts recommend that only the minimum number of power terms necessary to give an acceptable fit should be used. Of course, a perfect fit would be obtained if n data points were fitted with a power series in which the largest exponent was n − 1. However, the resulting equation could give strange values at points intermediate to those fitted. Fitting nonlinear expressions in MINITAB will be discussed at the end of the chapter.

If the values of the independent variable, X, are equispaced, multiple differences (i.e., differences of differences) may be used to indicate the maximum value of the exponent for the power series. Thus, if the third difference (difference of a difference of a difference) is approximately constant, the data could be appropriately represented by a second order equation.

Other Empirical Relationships

Some data are best represented by very complex expressions. There is ample theoretical evidence for such complex expressions in a number of cases and empirical relationships of this nature are also possible. Such expressions can be fitted using initial estimates of the parameters, obtained by graphical or other means. The expressions may be partially differentiated and higher derivatives are ignored. The first derivatives are linear and can be fitted by least squares as will be discussed below.

Fitting Data

Method of Selected Points

There are several ways to fit a set of data to an expression known or believed to be applicable. The simplest and least objective way is by the method of selected points. The data are first plotted and a graphical line is drawn to represent the points, based on judgment (eye estimate). In the case of a linear plot, the coordinates of two points on the line (X_1, Y_1, and X_2, Y_2), one near each extremity, are read and substituted in the general linear equation to obtain

$$Y_1 = a + bX_1$$
$$Y_2 = a + bX_2$$

where b is the slope of the line and a is the y intercept.

These equations are solved, simultaneously, to obtain values for the constants a and b. The values, Y_c, of this visually fitted line are computed for each value of X in the data set and compared with the observed values.

Values for the residuals, $r = Y_o - Y_c$, are computed and plotted with respect to the corresponding X values. A straight line, with ordinate of $r = 0$, is drawn parallel to the Y axis. A random scatter of the residual points around the line indicates a good fit of the data. See Chapter 9 for tests that can be used to decide on the randomness of residual plots.

A modification of the above method is first to average all of the X values and the Y values and to plot the single point, X, Y along with the values of X and Y. Draw a straight line through X, Y with the slope adjusted to what appears to best fit all of the plotted data. The coordinates of a lower and an upper point are read and substituted in the general equation as was done above. In many cases, such a plot is the "next best approximation" to a least squares fit, since the latter will always go through the point with coordinates X, Y.

Data that are nonlinear and cannot be linearized often can be fitted by an empirical power series, as discussed above. Again, the data are plotted and a smooth curve is drawn to provide a reasonable fit. Depending on the number of constants to be evaluated, the X and Y coordinates of points appropriately spaced are read off of the curve and substituted into the function. Thus, for a quadratic expression, one would obtain

$$Y_1 = a + bX_1 + cX_1^2$$
$$Y_2 = a + bX_2 + cX_2^2$$
$$Y_3 = a + bX_3 + cX_3^2$$

using selected points near the extremities and the middle of the plot. These equations are solved, simultaneously, to evaluate the constants a, b, and c. A residual plot is again constructed to ascertain goodness of fit.

It seems to be an unwritten law that one should use the minimum number of parameters to fit data. The best fit is achieved as the number of points is large with respect to the number of parameters used to fit the data.

Method of Averages

The method of averages is based on the assumption that a good fit is one in which the sum of the residuals is zero. In the case of a linear relationship, an equation of the type

$$Y_i = a + bX_i$$

is written for each pair of X_i, Y_i values. If these equations are listed in increasing order of the X values, the first half are summed to obtain an equation. The remaining half are summed in a like manner to obtain a second equation. These two summary equations are solved, simultaneously, to obtain an equation of a line that meets the assumed requirement, i.e., $\Sigma\, r = 0$.

Any combination of the individual equations may be used to obtain the summary equations and the groups combined do not need to be equal in number. Each combination so used will give a different equation and set of residuals. Even for a small number of data points, the number of possible combinations can be quite large. Thus, the number of different ways that $p + q$ different things can be divided into two groups is given by the expression

$$n = \frac{(p+q)!}{p!q!}$$

Example For five data sets, using combinations of two and three points for the equations

$$n = \frac{5!}{3!2!} = 10$$

However, the procedure described above (namely summing the first half and the second half of points) has been found by the author to be satisfactory on many occasions.

Example Consider the data set:

X	1.0	2.0	3.0	4.0	5.0	6.0	7.0	8.0	9.0
Y	12.5	19.0	32.0	37.5	48.2	62.5	72.0	79.0	92.5

The following equations may be written:

$$
\begin{array}{lr}
12.5 = a + 1.0\,b & (1)\\
19.0 = a + 2.0\,b & (2)\\
32.0 = a + 3.0\,b & (3)\\
37.5 = a + 4.0\,b & (4)\\
48.2 = a + 5.0\,b & (5)\\
62.5 = a + 6.0\,b & (6)\\
72.0 = a + 7.0\,b & (7)\\
79.0 = a + 8.0\,b & (8)\\
92.5 = a + 9.0\,b & (9)\\
\end{array}
$$

$\Sigma\,E$ 1 to 5 $149.2 = 5a + 15.0\,b$

$\Sigma\,E$ 6 to 9 $306.0 = 4a + 30.0\,b$

Solving the summary equations, simultaneously:

$$a = 1.267 \quad b = 10.369$$

The equation is

$$Y = 1.267 + 10.369\,X$$

The equation may be used to calculate values of Y_c for various values of X.

X	1.0	2.0	3.0	4.0	5.0	6.0	7.0	8.0	9.0
Y_o	12.5	19.0	32.0	37.5	48.2	62.5	72.0	9.0	92.5
Y_c	9.10	19.47	29.84	40.20	50.58	60.95	71.32	81.68	92.05
R	3.40	−.47	2.16	−2.70	−2.38	1.55	.68	−2.68	.45

$\Sigma r^2 = 36.57$

$$\Sigma r \,(\text{observed} - \text{computed}) = 0.01 \approx 0$$

Table 7.3. Normal Equations for Least Squares Curve Fitting for the General Power Series $Y = a + bX + cX^2 + dX^3 + ...$

$$\Sigma Y = na + \Sigma Xb + \Sigma X^2 c + \Sigma X^2 d + ...$$
$$\Sigma XY = \Sigma Xa + \Sigma X^2 b + \Sigma X^3 c + \Sigma X^4 d + ...$$
$$\Sigma X^2 Y = \Sigma X^2 a + \Sigma X^3 b + \Sigma X^4 c + \Sigma X^5 d + ...$$
$$\Sigma X^3 Y = \Sigma X^3 a + \Sigma X^4 b + \Sigma X^5 c + \Sigma X^6 d + ...$$

Table 7.4. Normal Equations for Least Squares Curve Fitting for the Linear Relationship $Y = a + bX$

$$\Sigma Y = na + \Sigma Xb$$
$$\Sigma XY = \Sigma Xa + \Sigma X^2 b$$

$$b = \frac{\Sigma XY - (X)(Y)/n}{\Sigma X^2 - \Sigma(X)^2/n}$$

$$a = \overline{Y} - b\overline{X}$$

Note that the sum of the residuals is zero (within rounding error) and that the sum of the first 5 residuals and also the sum of the remaining 4 residuals is zero, as well. These checks can be made to verify that the computations are correct.

A plot (not shown) of the residuals with respect to X appears to be random. The reader may wish to use other combinations of the data to obtain other fits to satisfy himself or herself as to how the method operates.

Nonlinear relationships may be fitted by the method of averages by first writing an equation for each set of data points and then forming summary equations in number equal to the parameters to be fitted. In the case of a quadratic equation, the first, middle, and upper third of the data may each be summed in a manner similar to what was described for the linear case. Other combinations could be used, of course.

Method of Least Squares

The method of least squares assumes that the best fit is one in which the sum of the squares of the residuals is a minimum. The classical way to achieve such a fit is to solve so-called normal equations that are formed according to the arrangement given in Table 7.3 for a power series (generalized parabola). The statistical literature generally refers to model fitting using the method of least squares as regression. Regression is a highly developed area of statistics and as such, most statistical software has facilities for performing regression. The interested reader can refer to [8] for an in-depth discussion of regression.

Table 7.4 contains the normal equations for the linear relationship which is, of course, a special case of the generalized parabola. The required summations may be computed from the respective values of X and Y to form the normal equations which are solved simultaneously.

Example Consider the previous data set.

X	Y	X^2	XY	Y^2
1.0	12.5	1	12.5	156.25
2.0	19.0	4	38.0	361.00
3.0	32.0	9	96.0	1024.00
4.0	37.5	16	150.0	1406.25
5.0	48.2	25	241.0	2323.24
6.0	62.5	36	375.0	3906.25
7.0	72.0	49	504.0	5184.00
8.0	79.0	64	632.0	6241.00
9.0	92.5	81	832.5	8556.25
Σ 45.0	455.2	285	2881.0	29158.24

From Table 7.3, the normal equations for a linear relationship are:

$$\Sigma Y = n\,a + \Sigma X\,b$$
$$\Sigma XY = \Sigma X\,a + \Sigma X^2\,b$$

Substituting numerical values

$$455.2 = 9\,a + 45\,b$$
$$2881.0 = 45\,a + 285\,b$$

These equations may be solved, simultaneously, to obtain

$$a = 0.161 \quad b = 10.083$$

so that the equation becomes

$$Y = 0.161 + 10.083 \, X$$

This equation may be used to calculate values for Y_c for the nine values of X, as follows:

X	1.0	2.0	3.0	4.0	5.0	6.0	7.0	8.0	9.0
Y_o	12.5	19.0	32.0	37.5	48.2	62.5	72.0	79.0	92.5
Y_c	10.24	20.33	30.41	40.49	50.58	60.66	70.74	80.83	90.91
r	2.26	−1.33	1.59	−2.99	−2.38	1.84	1.26	−1.83	1.59
r^2	5.11	1.77	2.53	8.94	5.66	3.39	1.59	3.35	2.53

$\Sigma r^2 = 34.87$

Note that this sum is smaller than that obtained in the example for the method of averages, namely $\Sigma r^2 = 36.57$.

Table 7.5. Basic Worksheet for All Types of Linear Relationships

X denotes _____   Y denotes _____

$\Sigma X =$ _____ $\Sigma Y =$ _____

$\overline{X} =$ _____ $\overline{Y} =$ _____

Number of points: n   =   _____

Step   (1) ΣXY =   _____

(2) $(\Sigma X)(\Sigma Y)/n$ =   _____

(3) S_{xy} =   Step(1) − Step(2)

(4) ΣX^2 =   _____   (7) ΣY^2 =   _____

(5) $(\Sigma X)^2/n$ =   _____   (8) $(\Sigma Y)^2/n$ =   _____

(6) S_{xx} =   Step(4) - Step(5)   (9) S_{yy} =   Step(7)-Step(8)

(10) $b = S_{xy}/S_{xx}$ =   Step(3) / Step(6)

(11) Y   =   _____   (12) $b\overline{X} =$

(13) $a = Y - b\overline{X}$ =   Step (11) - Step (12)

(14) $(S_{xy})^2/S_{xx}$ =   _____

(15) $(n-2) s_Y^2$ =   Step (9) - Step (14)

(16) s_Y^2 =   Step(15)/(n-2)

(17) $s_Y = \sqrt{}$ (Step 16)   =   _____

(18) $s_b^2 = s_Y^2 / S_{xx} = $ _____Step(16)/Step(6)_____

(19) $s_a^2 = s_Y^2 [(\frac{1}{n}) + (\frac{\bar{X}^2}{S_{xx}})] = $ _____

Equation of the line $Y = a + bX$

*The following are algebraically identical:

$$S_{xx} = \Sigma(X-\bar{X})^2; S_{yy} = \Sigma(Y-\bar{Y})^2; S_{xy} = \Sigma(X-\bar{X})(Y-\bar{Y})$$

Caution: In hand calculating, carry all decimal places obtainable, i.e., if data are recorded to two decimal places, carry four decimal places in the computations, to avoid losing significant figures in subtraction.

Taken from: M.G. Natrella, "Experimental Statistics" [9].

Table 7.5 [9] consists of a format that may be used for linear least squares fitting of data. It includes, as additional steps, the calculations to compute the estimated variances of the slope and the intercept as well as the equation of the line. Many statistical software packages contain programs to compute these quantities as well. *Example* Consider the data used previously. The sums of the squares and the cross products have been calculated, already, so will be used here as well.

	$\bar{X} = 5.0$		$\bar{Y} = 50.577778$
(1)	$\Sigma XY = 2881$		
(2)	$(\Sigma X)(\Sigma Y)/n = 2276$		
(3)	$S_{xy} = 605$		
(4)	$\Sigma X^2 = 285$	(7)	$\Sigma Y^2 = 29158.24$
(5)	$(\Sigma X)^2/n = 225$	(8)	$(\Sigma Y)^2/n = 23023.00$
(6)	$S_{xx} = 60$	(9)	$S_{yy} = 6135.24$
(10)	$S_{xy}/S_{xx} = b = 10.0833$	(14)	$(S_{xy})^2/S_{xx} = 6100.4167$
(11)	$\bar{Y} = 50.5778$	(15)	$(n-2)s_Y^2 = 34.8233$
(12)	$= 50.4167$	(16)	$s_Y^2 = 4.9748$
(13)	$a = 0.1611$	(17)	$s_Y = 2.2304$

Estimated variance of the slope, b

(18) $$s_b^2 = s_Y^2 / S_{xx} = .082913, \quad s_b = 0.2879$$

Estimated variance of the intercept, a

(19) $$s_a^2 = s_Y^2 [1/n = X^2 / S_{xx}] = 2.6256, \quad sa = 1.626$$

The equation of the line is:

$$Y = 0.161 + 10.0833$$

This is identical with the equation calculated above by the alternative method. Accordingly, the residuals are the same as well.

MINITAB Example

Consider the MINITAB dataset 'furnace.mtw'. For 90 gas-heated homes, the average energy consumption with the damper in (c8 BTU In) and the age of the home (c7 Age) was recorded. Other information on the homes was collected but not shown below. The first few observations in the dataset are shown in Table 7.6.

Table 7.6.   Furnace Dataset from MINITAB

Age	BTU In
8	7.87
75	9.43
44	7.16

Taken from MINITAB

The data will be fit using regression with the method of least squares and the residuals examined for significant model departure. The following MINITAB menu commands fit the regression model:

Under Stat, Regression, Regression
- double click C8 BTU In to add it to the **Response** field
- double click C7 Age to add it to the **Predictors** field
- under **Graphs, Residual Plots**, select Normal Plot of Residuals, click OK, click OK

The MINITAB output is shown with the normal probability plot in Figure 7.6.

Since the normal probability plot indicates a fairly straight line, no systematic departures from the model are suspected. The equation of the line is given by

$$y = 9.52 + 0.0135x$$

or

$$BTUin = 9.52 + 0.0135age.$$

MINITAB Output

```
Regression Analysis: BTU.In versus Age

The regression equation is
BTU.In = 9.52 + 0.0135 Age

Predictor           Coef      SE Coef          T        P
Constant          9.5181       0.4808      19.80    0.000
Age             0.013492     0.009727       1.39    0.169
S = 2.853         R-Sq = 2.1%      R-Sq(adj) = 1.0%
Analysis of Variance
Source               DF           SS          MS        F        P
Regression            1       15.662      15.662     1.92    0.169
Residual Error       88      716.395       8.141
Total                89      732.058

Unusual Observations
Obs    Age    BTU.In         Fit      SE Fit     Residual    St Resid
  7   45.0    16.900      10.125       0.307        6.775        2.39R
 20   60.0     4.000      10.328       0.366       -6.328       -2.24R
 37   80.0    18.260      10.597       0.503        7.663        2.73R
 43    2.0     2.970       9.545       0.466       -6.575       -2.34R
 50   50.0    16.060      10.193       0.321        5.867        2.07R
R denotes an observation with a large standardized residual
```

Taken from MINITAB

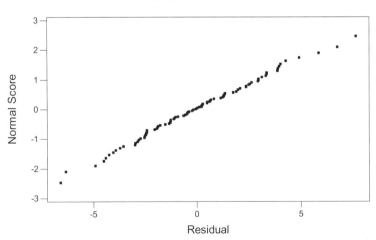

Normal Probability Plot of the Residuals

(response is BTU.In)

Figure 7.6.   Normal probability plot of residuals.

One useful feature of the MINITAB output provided is that it gives a listing of any observations considered outliers due to their large standardized residuals. Although this is not pursued further for the furnace data, each complete data set analysis requires further investigation of such information. Note that age is not a significant predictor of energy use (as indicated by comparing T=1.39 to T_c=1.96). It is possible to fit a nonlinear model of the form

$$y = a + bx + cx^2$$

to the furnace data. With MINITAB the C7 age variable can easily be manipulated to calculate C11 agesquare:

Under Calc, Calculator
- enter C11 under the **Store Result In Variable** field
- enter C7 Age **2 under the **Expression** field, click OK

This creates the variable C11 which is the square of C7. The nonlinear regression menu commands are then

Under Stat, Regression, Regression
- double click C8 BTU In to add it to the **Response** field
- double click C7 Age and C11 Agesquare to add them in the **Predictors** field, click OK

The resulting MINITAB output is shown below:

MINITAB Output

```
Regression Analysis: BTU.In versus Age, Agesquare

The regression equation is
BTU.In = 7.78 + 0.144 Age - 0.00134 Agesquare

Predictor        Coef      SE Coef           T          P
Constant       7.7788       0.6231       12.48      0.000
Age           0.14352      0.03385        4.24      0.000
Agesquar   -0.0013405    0.0003364       -3.98      0.000
S = 2.639        R-Sq = 17.2%     R-Sq(adj) = 15.3%


Analysis of Variance
Source             DF           SS           MS        F         P
Regression          2      126.222       63.111     9.06     0.000
Residual Error     87      605.836        6.964
Total              89      732.058


Source         DF       Seq SS
Age             1       15.662
Agesquar        1      110.559
```

```
Unusual Observations
Obs    Age      BTU.In        Fit      SE Fit     Residual    St Resid
  7    45.0     16.900     11.523      0.451        5.377       2.07R
 20    60.0      4.000     11.564      0.459       -7.564      -2.91R
 37    80.0     18.260     10.681      0.466        7.579       2.92R
 67    99.0     15.120      8.849      0.791        6.271       2.49R
R denotes an observation with a large standardized residual
```
Taken from MINITAB

The nonlinear model including the square of age is now

$$BTUin = 7.78 + 0.144age - 0.00134age^2.$$

Note that the T values for age and the square of age both fall within the critical rejection region. This is not uncommon when modeling age [8]. The reader should verify that the normal probability plot of the residuals is a fairly straight line.

Summary

The history of the development and the general acceptance of the method of least squares is discussed in an interesting paper by Eisenhart [10]. A least squares fit is essentially the most objective fit that one can obtain, since no judgment of the data analyst is involved. It is accepted to be the best fit when the law of error is Gaussian. In other cases it can be depended on to yield nearly-best estimates when the number of independent observations is large with respect to the number of parameters to be determined [9].

With the risk of being repetitious, it will be emphasized that the data points used must be independent to provide a good representation of the subject to which they relate. This is true no matter what method of fit is used. When the method of least squares is used, it is especially important that erroneous (or outlying) data points not be included in the data set fitted. The eye will give little weight and even ignore points far removed from a line in the case of a graphical fit. The method of least squares assumes that all points are equally meritorious and will skew a line, if necessary, to minimize the sum of the squares. Such a displacement will be obvious when a residual plot is made when it may be necessary to discard the culprit and refit the line. Initial plotting was a rule followed by data analysts when least squares fits were made, laboriously, using desk calculators. It is still a good idea. The method of the "huge error" described in Chapter 6 may be used to decide when to discard a point as a result of an initial plot or a residual plot.

EXERCISES

7-1. Given the data set

X	1	2	3	4	5	6
Y	11	18	29	43	49	61

Fit the function $Y = a + bX$ by the method of selected points.

7-2. Fit the data of exercise 7-1 using the method of averages.

7-3. Fit the data of exercise 7-1 using the method of least squares. Perform this calculation using MINITAB.

7-4. Estimate the standard deviation of the slope and of the intercept for the equation fitted in exercise 7-3. Do you consider the slope and or the intercept to be significant? Answer the same question using MINITAB.

7-5. Given the data set

X	1	2	3	4	5	6	7	8	10
Y	2.0	1.50	1.33	1.25	1.20	1.17	1.14	1.125	1.00

Plot the original data. Transform the data to linearize it and plot the transformed data. Fit a linear equation to the transformed data and derive the equation for the original data.

7-6. Given the data set

X	1	2	3	4	5	6	7	8	9	10
Y	3.0	5.0	8.0	11.0	13.5	16.5	20.0	23.5	27.2	31.0

Fit the data to the equation

$$Y = a + bX + cX^2$$

by the method of averages. Make a residual plot and calculate the sum of the residuals and the sum of the squares of the residuals. Repeat using MINITAB.

7-7. Fit the data set of exercise 7-6 to the equation

$$Y = a + bX + cX^2$$

by the method of least squares. Make a residual plot and calculate the sum of the squares of the residuals.

REFERENCES

[1]   Lipka, J. *Graphical and Mechanical Computation* (New York: John Wiley & Sons, 1918).

[2]   J.B. Scarborough, "Numerical Mathematical Analysis," 6th Edition, Johns Hopkins University Press, Baltimore, MD (1966).

[3]   Lotus Development Corporation, 55 Cambridge Parkway, Cambridge, MA 02142.

[4]   *ASTM Manual on Presentation of Data and Control Chart Analysis* (Philadelphia: ASTM).

[5]   Cleveland W.S. *The Elements of Graphing Data* (Monterey, CA: Wadsworth Advanced Books and Software, 1985).

[6]   L.P. Provost, "Statistical Methods in Environmental Sampling," in *Environmental Sampling for Hazardous Waste*, ACS Symposium Series 267, (Washington DC: American Chemical Society, 1984).

[7]   Northwest Analytical, Inc., 520 NW Davis, Portland OR 97209.

[8]   Neter, J. Wasserman, W. Kutner, M. *Applied Linear Statistical Models 3rd Edition* (Homewood, IL: Irwin, 1990).

[9]   Natrella M.G, *Experimental Statistics* NBS Handbook 91, (Gaithersburg, MD: National Institute of Standards and Technology).

[10] Eisenhart C., "The Meaning of 'Least' in Least Squares," *J. Wash. Acad. Sci.,* 54: 24-33 (1964).

CHAPTER 8

Proportions, Survival Data, and Time Series Data

This chapter introduces proportions, survival data, and time series data to the reader. The statistical methods developed thus far have been appropriate for continuous and in particular, normally distributed data. Statistical analyses for other commonly encountered data types are discussed and computer techniques for each data type are evaluated.

INTRODUCTION

The material in the preceeding chapters was largely focused on normally distributed data. Statistical techniques were presented for assessing normality of data and describing the data. Confidence intervals were summarized for one-sample and two-sample situations. The data sets discussed included radiation measurements taken using different methods (filter, membrane, open cup, and badge) or percentage of sulphur in fuel oil. Although continuous data is frequently encountered, other types of data (and as such other types of data analyses) are arising more frequently in metrology.

Proportion data is based on the number of items displaying a certain outcome. For example, a number of items may be tested and only some of them may be considered acceptable. Statistical methodology for this situation is focused on describing and making inferences regarding the proportion of acceptable items.

Survival data summarizes the time of a particular event. Rather than obtaining information concerning whether an item was considered acceptable or not, the items may be monitored to determine the exact time at which each one fails (and hence become unacceptable). For this reason, survival data is often referred to as failure time data. Since all items being monitored may not fail during the fixed monitoring time, not all items will have an associated failure time and hence may be censored.

Time series data summarizes a measurement at various points thoughout time. One may measure the percentage of sulphur in gasoline on a weekly basis thoughout the year.

Each of these types of data can be highly useful in certain situations, and as such, have many statistical books and journals devoted solely to their specific analysis methods. The information presented here is meant to be a general overview of these types of data and simple statistical analyses that can be performed on them with the assistance of computer software. Statistical tests for frequently encountered situations are discussed but the material presented is by no means a comprehensive summary of the analyses available. The reader is urged to examine the reference material for a more sophisticated statistical discussion on the subject matter.

PROPORTIONS

Introduction

The evaluation of proportions such as the fraction of acceptable items in a population is of some interest to metrologists from several points of view. There is an interest in the confidence limits for a measured proportion. A related question is the number of individuals that need to be examined to establish a proportion within a specified limit of uncertainty. Then, questions of the significance of apparent differences of measured proportions or ratios can be of interest.

Consider the measurement of some property from a random sample of a population to decide on its acceptability with respect to a desired outcome. An example would be the number of defective items in a lot of material.

Let   n = number of items sampled
       r = number of items that are acceptable in the sample
       p = r/n = the proportion of acceptable items in the sample

The best estimate of the proportion of acceptable items is p, the measured value.

The text below discusses one sample and two sample confidence intervals, both one-sided and two-sided. A brief discussion of sample size estimation is also given.

One Sample Topics

It is obvious that if another sample of items was taken, some different value for p could be obtained. How does a measured value for p correspond to the true proportion, P, of acceptable items in the lot? This cannot be answered unless all items in the lot were tested. However, a confidence interval can be calculated (as in the case of a mean already discussed) that is expected to include the true value, P, of the proportion with a stated confidence. A two-sided confidence interval gives upper and lower limits for the proportion. One-sided intervals also can be calculated to indicate that P is not expected to exceed or conversely, a value that P is expected not to be less than.

The method used to calculate confidence intervals for small samples (i.e., small number of items tested, <30) involves the binomial distribution and is beyond the scope of the present book. Reference 1 gives more information on using the binomial distribution for small samples. In addition, tables exist that can be consulted to obtain this information. Table A-23, A-24, and A-25 of Reference 2 are examples.

Two-Sided Confidence Intervals for One Sample

A confidence interval (p_l, p_u) around the sample proportion p may be calculated using the following formula:

Confidence Interval	$p_l = p - z_{1-\alpha/2} \sqrt{[p(1-p)/n]}$
for p (two-sided)	$p_u = p + z_{1-\alpha/2} \sqrt{[p(1-p)/n]}$

This confidence interval is appropriate as long as the sample size is larger than 30.

Questions as to whether a measured proportion differs significantly from an expected or required value for a proportion are answered with reference to a calculated confidence interval. In the case of an expected or required value, if it lies within the confidence interval, there is no reason to believe (on the basis of the experimental evidence) that the measured value differs significantly from that expected or required.

Example   The following illustrates calculation of a two-sided confidence interval for a proportion.

Consider the testing of a material in which there needs to be a 95% confidence that 95% of the items are acceptable. A sample of 50 items was randomly selected and 45 of them were acceptable. Does the material meet the requirements for acceptability?

$$n = 50, r = 45, p = 0.90, z_{1-\alpha/2} = 1.96$$
$$p_l = .90 - 1.96\sqrt{[.90(.10)/50]} = 0.8168$$
$$p_u = .90 + 1.96\sqrt{[.90(.10)/50]} = 0.9831$$

Since the required proportion, 0.95, is inside of the 95% confidence interval (0.82,0.98), the material is considered to meet the specification.

MINITAB Example

The following MINITAB menu commands calculate the same confidence interval:

Under Stat, Basic Statistics, 1 Proportion
- select **Summarized Data**, enter 50 under **Number of Trials**, 45 under **Number of Events**
- under **Options**, enter 95.0 under **Confidence Interval**, not equal to under **Alternative**
- select **Use Test and Interval Based on Normal Distribution**, click OK, click OK

The following MINITAB output is produced:

MINITAB Output

```
Test and CI for One Proportion

Test of p = 0.5 vs p not = 0.5

Sample     X     N   Sample p        95.0% CI          Z-Value   P-Value
1         45    50   0.900000   (0.816846, 0.983154)    5.66     0.000

* NOTE * The normal approximation may be inaccurate for small samples.
```

Taken from MINITAB

One Sided Confidence Intervals for One Sample

The one-sided lower confidence interval (p_l, 1) and one-sided upper confidence interval (0,p_u) may be calculated using the following formulas:

Confidence Interval	$p_l = p - z_{1-\alpha} \sqrt{[p(1-p)/n]}$
for p (one-sided)	$p_u = p + z_{1-\alpha} \sqrt{[p(1-p)/n]}$

The only difference between the one-sided and two-sided formulas is that one-sided formulas have all the error α in the one side of the confidence interval that is of most interest whereas two-sided confidence intervals split the error equally between both sides, $\alpha/2$.

Example Referring back to the previous example, suppose one now wants to know the proportion for which there will be 95% confidence that the true proportion will not be less than. Simply stated, this means the lower bound on the proportion of acceptable items. Using the formula above for p_l gives

$$p_l = .9 - 1.645\sqrt{[.9(.1)/50]} = 0.8302$$

so that the 95% lower confidence interval is (0.83,1). This means we have 95% certainty that the minimum proportion of acceptable items is 83%.

MINITAB Example

The one-sided confidence interval is calculated in MINITAB using the following menu commands:

Under Stat, Basic Statistics, 1 Proportion
- select **Summarized Data**, enter 50 under **Number of Trials**, 45 under **Number of Events**
- under **Options**, enter 95.0 under **Confidence Interval**, greater than under **Alternative**
- select **Use Test and Interval Based on Normal Distribution**, click OK, click OK

The following MINITAB output is produced:

MINITAB Output

```
Test and CI for One Proportion

Test of p = 0.5 vs p > 0.5

Sample      X      N   Sample p   95.0% Lower Bound   Z-Value   P-Value
1          45     50   0.900000          0.830215      5.66     0.000

* NOTE * The normal approximation may be inaccurate for small samples.
```

Taken from MINITAB

Example Opinion polls may be of interest to some readers. Since the standard deviation of a proportion is

$$\sqrt{[p(1-p)/n]}$$

one can calculate the statistical confidence interval (95% confidence) of a poll proportion, p, based on n votes as

$$2\sqrt{[p(1-p)/n]}$$

This is based on a random sampling of the population which may be difficult to prove. It is obvious that only a biased opinion can come from a biased sample, and this must always be considered to be possible, even in the best designed polls.

Sample Sizes for Proportions-One Sample

The alert reader may observe that all one has to do to make a material pass is to use a smaller sample! This is because the confidence limits become wider as the sample size is decreased, reflecting the fact that a smaller sample results in a more variable estimate. To prevent such a misuse of statistics, a minimum sample size should be specified a priori that will limit the error of estimation to an acceptable level. This may be calculated for a two-sided test by the expression

$$n = \left(\frac{z_{1-\alpha/2}\sqrt{p_0(1-p_0)} + z_\beta\sqrt{p_1(1-p_1)}}{p_1 - p_0} \right)^2$$

where

p_0 = proportion under the null hypothesis (acceptable)

p_1 = proportion under the alternative hypothesis (unacceptable)

For a one-sided test, use $z_{1-\alpha}$ as mentioned before.

Most experiments that are properly designed first involve a sample size calculation to determine how many experimental units should be studied for the experimenter to draw conclusions with reasonable certainty.

Suppose for the previous example that one would like to know how many items should be sampled to detect a 4% decrease in the number of acceptable items. The acceptability rate is set at 94%, and the required power is 99% with significance level 5%. The required sample size for a one-sided test would then be

$$n = \left(\frac{1.645\sqrt{.94(.06)} + 2.33\sqrt{.9(.1)}}{.04} \right)^2$$

$$= 741$$

MINITAB Example

The sample size may be found in MINITAB using the following menu commands:

Under Stat, Power and Sample Size, 1 Proportion
- enter 0.9 under **Alternative Values for P**
- enter 0.99 under **Power Values**
- enter 0.94 under **Hypothesized P**
- under **Options, Alternative Hypothesis**, select less than, under **Significance Level** enter 0.05, click OK, click OK

The output produced in MINITAB is shown below:

MINITAB Output

```
Power and Sample Size

Test for One Proportion

Testing proportion = 0.94 (versus < 0.94)
Alpha = 0.05

  Alternative   Sample   Target   Actual
  Proportion     Size    Power    Power
    0.900000       741   0.9900   0.9900
```

Taken from MINITAB

Two Sample Topics

One sample proportion questions involve inferences surrounding one population. The metrologist is interested in how the sample proportion is behaving. It is intuitative that two sample proportion questions involve inferences surrounding two populations. In this case the focus is on how the two sample proportions behave in relation to each other. For example, a lot of items may be manufactured using a different process than the standard. It is of interest to compare the proportion of defective items manufactured under the old and new processes. This is generally done by examining the difference between the proportions from the two samples, $d=p_1-p_2$.

Let n_1= number of items sampled in the new lot
  r_1= number of items that are acceptable in the new lot sample
  $p_1=r_1/n_1$ the proportion of acceptable items in the new lot sample.

Similarly, let
    n_2 = number of items sampled in the standard lot
    r_2= number of items that are acceptable in the standard lot sample

$p_2 = r_2/n_2 =$ the proportion of acceptable items in the standard lot sample.

The best estimate of the proportion of acceptable items in sample 1 and 2 are p_1 and p_2, the measured values.

Two Sided Confidence Intervals for Two Samples

A two sided confidence interval (d_l, d_u) around the difference between two sample proportions is calculated as

Confidence Inverval	$d_l = p_1 - p_2 - z_{1-\alpha/2} \sqrt{\dfrac{(r_1 + r_2)}{(n_1 + n_2)} \left(1 - \dfrac{(r_1 + r_2)}{(n_1 + n_2)}\right) \left(\dfrac{1}{n_1} + \dfrac{1}{n_2}\right)}$
for d	$d_u = p_1 - p_2 + z_{1-\alpha/2} \sqrt{\dfrac{(r_1 + r_2)}{(n_1 + n_2)} \left(1 - \dfrac{(r_1 + r_2)}{(n_1 + n_2)}\right) \left(\dfrac{1}{n_1} + \dfrac{1}{n_2}\right)}$

Note that the formula above uses a pooled estimate of the variability.
Example   A metrologist examines the accuracy of measurements taken using two devices. The metrologist finds that 90 of the 100 measurements taken using device one were accurate versus 94 of the 100 measurements taken using device two. Can the metrologist conclude with 95% certainty that the accuracy of the two devices is the same? The computations are as follows:

$$p_1 = \frac{90}{100} = 0.9 \quad p_2 = \frac{94}{100} = 0.94$$

$$d_l = 0.9 - 0.94 - 1.96 \sqrt{\frac{(90+94)}{(100+100)} \left(1 - \frac{(90+94)}{(100+100)}\right) \left(\frac{1}{100} + \frac{1}{100}\right)} = -0.115$$

$$d_u = 0.9 - 0.94 + 1.96 \sqrt{\frac{(90+94)}{(100+100)} \left(1 - \frac{(90+94)}{(100+100)}\right) \left(\frac{1}{100} + \frac{1}{100}\right)} = 0.035$$

Since the confidence interval (-0.115, 0.035) includes zero, the metrologist concludes with 95% certainty that there is no difference in the accuracy of the two devices.

MINITAB Example

To perform this type of analysis on two proportion data using MINITAB, select the following menu commands which yield the subsequent output:

Under Stat, Basic Statistics, 2 Proportions

- select **Summarized Data**
    - enter 100 under **First Sample Trials**, 90 under **Events**
    - enter 100 under **Second Sample Trials**, 94 under **Events**
- under Options, select **Use Pooled Estimate of p for Test**, click OK, click OK

MINITAB Output

```
Test and CI for Two Proportions
Sample       X      N   Sample p
1           90    100   0.900000
2           94    100   0.940000

Estimate for p(1) - p(2):   -0.04
95% CI for p(1) - p(2):   (-0.114993, 0.0349926)
Test for p(1) - p(2) = 0 (vs not = 0):   Z = -1.04   P-Value = 0.297
```
Taken from MINITAB

Chi-Square Test of Association

As in many statistical problems, the question as to whether two proportions differ may be answered in another fashion. Rather than comparing the two proportions (as was done with the confidence interval around the difference in proportions), the frequency counts that gave rise to the proportions may be examined. For this purpose, it is helpful to summarize the data in the following fashion:

		Outcome		
		Accurate	Not Accurate	Row total
Device	1	90	10	100
	2	94	6	100
	Column Total	184	16	200

Or similarly,

		Variable 2		
		Outcome 1	Outcome 2	Row total
Variable 1	1	A	B	A+B
	2	C	D	C+D
	Column Total	A+C	B+D	N

This is referred to as a 2x2 table. An association between device and accuracy can be tested using the Chi-Square statistic, given by

$$\chi^2 = \frac{(N-1)(AD-BC)^2}{(A+C)(B+D)(A+B)(C+D)}.$$

The above test statistic is appropriate for large samples (both greater than 20). The calculation procedure follows:

1. Choose α, the significance level of the test.

2. Compute χ^2 as above.

3. Look up the value of $\chi^2_{1-\alpha}$ for 1 degree of freedom from the following table :

$\chi^2_{.90}$	$\chi^2_{.95}$	$\chi^2_{.975}$	$\chi^2_{.99}$	$\chi^2_{.995}$
2.71	3.84	5.02	6.63	7.88

4. If $\chi^2 > \chi^2_{1-\alpha}$, decide that the two proportions differ, otherwise the experimental evidence does not support such a decision.

With level of significance 0.05 (95% certainty), the device accuracy computation is

$$\chi^2 = \frac{(199)\left((90)(6)-(94)(10)\right)^2}{(184)(16)(100)(100)}$$
$$= 1.08$$

Comparing this value to the Chi Square value of 3.84 at level of significance 0.05, one can see that the experimental evidence does not support the proportions differing.

MINITAB Example

To perform this calculation using MINITAB, enter the data in the spreadsheet portion as shown:

C1	C2
90	10
94	6

Then select the following menu commands:

Under Stat, tables, Chi-square test
- double click on C1 and C2 to add them under **Columns Containing the Table**, click OK

The following MINITAB output is produced:

MINITAB Output

```
Chi-Square Test: C1, C2

Expected counts are printed below observed counts
            C1        C2      Total
    1       90        10       100
          92.00      8.00
    2       94         6       100
          92.00      8.00

Total      184        16       200
Chi-Sq =   0.043 +   0.500 + 0.043 +   0.500 = 1.087
DF = 1, P-Value = 0.297
```

Taken from MINITAB

The bold portion of the MINITAB output re-confirms the chi-square statistic value of 1.08.

Note that extensions of the statistic are available for testing associations between variables with more than two levels [3]. Procedures useful for this purpose are also described by Natrella [2].

One Sided Confidence Intervals for Two Samples

The one-sided lower confidence interval $(d_l,1)$ and one-sided upper confidence interval $(0,d_u)$ around the difference in two proportions are given by

$$d_l = p_1 - p_2 - z_{1-\alpha}\sqrt{\frac{(r_1+r_2)}{(n_1+n_2)}\left(1-\frac{(r_1+r_2)}{(n_1+n_2)}\right)\left(\frac{1}{n_1}+\frac{1}{n_2}\right)}.$$

$$d_u = p_1 - p_2 + z_{1-\alpha}\sqrt{\frac{(r_1+r_2)}{(n_1+n_2)}\left(1-\frac{(r_1+r_2)}{(n_1+n_2)}\right)\left(\frac{1}{n_1}+\frac{1}{n_2}\right)}$$

Sample Sizes for Proportions-Two Samples

As with the one sample case, it is possible to calculate the required sample size to test for a specified difference between two groups with a certain power and error rate. The sample size per group for a two-sided test is given by

$$n = \left(\frac{z_{1-\alpha/2}\sqrt{(p_1 + p_2)(q_1 + q_2)/2} + z_{1-\beta}\sqrt{p_1q_1 + p_2q_2}}{(p_1 - p_2)} \right)^2$$

where $q_1 = 1 - p_1$ and $q_2 = 1 - p_2$. For a one-sided test, use $z_{1-\alpha}$. Suppose for the previous example, the metrologist would like to know with 95% confidence how big a sample is required to test with 90% power differences greater than 5%, assuming the device accuracy rate is 92% and only situations where device two is less accurate than device one are of interest (i.e., a one-sided test). This may be calculated as

$$n = \left(\frac{1.645\sqrt{(.92 + .87)(.08 + .13)/2} + 1.28\sqrt{(.92)(.08) + (.87)(.13)}}{(.92 - .87)} \right)^2$$

$$= 641.33$$

$$= 642$$

MINITAB Example

The following MINITAB menu commands calculate the required sample size:

Under Stat, Power and Sample Size, 2 Proportions
- enter 0.92 under **Proportion 1 Values**
- enter 0.9 under **Power Values**
- enter 0.87 under **Proportion 2**
- under **Options, Alternative Hypothesis**, select Greater Than, under **Significance Level**, enter 0.05, click OK, click OK

The output is shown below.

MINITAB Output

```
Power and Sample Size

Test for Two Proportions
Testing proportion 1 = proportion 2 (versus >)
Calculating power for proportion 2 = 0.87
Alpha = 0.05

                Sample   Target   Actual
Proportion 1     Size    Power    Power
   0.920000       642   0.9000   0.9000
```

Taken from MINITAB Output

SURVIVAL DATA

Introduction

Survival data has several features that make it unlike any other type of data. Knowing the actual time an event occurs provides much more information than knowledge of whether or not an event occurs (as with proportion data). It is much more informative to say 'among the 10% of units that failed, the mean failure time was 136 hours' than to say '10% of units failed'. This extra information intuitively makes survival data more interesting. The broadly used term 'survival data' actually refers to the time of any event. Examples can be found in diverse areas where statistics are applied: time of AIDs positivity in healthcare data, time of circuit failure in engineering data, or time of engine failure in metrology data.

Censoring

Implicit in survival data is the fact that the event of interest may not occur. If the event does occur, one records the event time and the event is considered uncensored. If the event does not occur, the time of event is recorded as the last time on study and the event is considered censored. The result is a time outcome (t) and an indicator variable for censoring (δ). An important assumption for survival data analysis is referred to as *non-informative censoring*: t, the actual survival time, is independent of any mechanism causing censoring at time c, c<t. For example, non-informative censoring is violated in the healthcare data example if patients on one toxic AIDs treatment drop out early due to side effects (and thus are censored) if that group would have experienced earlier time of AIDs positivity had they remained in the study.

There are several types of censoring that may occur in survival data:

Interval Censoring: occurs when the true event time is between two times, eg., batteries are changed in a flashlight, check the flashlight every three months for battery power. If the flashlight lights at 3 months but not at 6 months, the battery failure time is interval censored. We know the battery failure time was after 3 months but before 6 months. Interval censored data implies periodic monitoring of whether or not an event occurred.

Left Censoring: occurs when the true event time is less than the censored event time, eg., a tumor is removed from cancer patients, check the patients 3 months later for tumor recurrence. If patient has tumor recurrence, he or she is left censored. The time of recurrence becomes 3 months (even though true time was before). Left censored data implies the event of interest has occurred before the monitoring of the event time even started.

Right Censoring: occurs when the true event time is greater than the censored event time, eg., observe engaged couples for time of marriage in sociological study. If the couple does not marry during the study, they are right censored. The time of marriage is recorded as the study end time and their outcome is considered censored. Right censored data implies the event of interest occurs sometime after the monitoring of the event time stopped.

Within right censored data, there are several types of studies that may be conducted:

Type I – The study units are observed for a *fixed period* of time, and the exact failure times are recorded for those failing during observation period. Those not failing are considered censored, with failure times set at time of last observation/end of study.

Type II – The study units are observed until a *fixed number* of failures occur. The failure times for censored/uncensored units are the same as in Type I censoring.

Random – The study units are not all available for monitoring at the start of the study, so that the units may have unequal times under study. The failure times reflect the period under study for that unit. Failure times for censored/uncensored units are the same as in Type I censoring. This type of data is most commonly encountered in situations where units are accrued gradually over time.

Right censored Type I methods will be discussed here due to general applicability in metrology. Other methods and/or modifications of methods exist for the remaining situations.

One Sample Topics

The analysis of survival data is primarily focused on estimating the survivor function, $S(t)$, which gives the probability of survival at time t. Possible survival function estimation methods include the Product Limit/Kaplan Meier and Life Table methods. The Product Limit/Kaplan Meier method calculates survival estimates at each event time, whereas the Life Table method calculates survival estimates during specified time intervals (ignoring the exact event times). The life table method only uses the number of events and the number of units censored during a given interval for calculations, making it an ideal estimate of survival if the data is interval censored. If the actual event times are known, the Product Limit/Kaplan Meier method is superior because it makes use of all information present and results in a more accurate estimate of survival. Note that survival functions may be estimated within one group or for several groups and then compared to see if the groups differ with respect to estimated survival.

Product Limit/Kaplan Meier Survival Estimate

Assume we have failure times, $t_1, t_2, ..., t_r$. The formula for the Product Limit/Kaplan Meier survival estimate at a given time, t_k is

$$S(t_k) = \prod_{j=1}^{k} \frac{n_j - d_j}{n_j}$$

The procedure to be used in calculating the Product Limit/Kaplan Meier survival estimate is as follows:

1. Order the event times in ascending order, we will call these times 1,2,...,r.
2. Calculate

$$\frac{n_j - d_j}{n_j}$$

at each of the event times, where

n_j=number not failing (at risk of failure) just before the time of the jth failure, t_j
d_j=number of failures at t_j (1 if no ties)

3. Calculate

$$S(t_k) = \prod_{j=1}^{k} \frac{n_j - d_j}{n_j}$$

over all event times. Although the symbol

$$\prod_{j=1}^{k}$$

may appear intimidating, it simply means multiply.

Note that by definition $S(t)=1$ for t< t_1 , i.e., probability of not failing is 100% until the first unit fails. If the last failure time is censored, $S(t)$ will not be defined beyond that time.

As with many statistical tests, the above test has two names: the Product Limit and the Kaplan Meier survival estimate. MINITAB refers to it as the Kaplan Meier estimate and so will our text from this point forward.

MINITAB Example

Example Consider the MINITAB dataset 'reliable.mtw'. This data gives time of failure for engine windings in a turbine assembly. Fifty windings were tested at 80 degrees C; 40 windings were tested at 100 degrees C. Some units were removed from the test due to failures from other causes. Censoring indicators are used to designate which times are actual failures (indicator = 1) and which are units removed from the test (indicator = 0). The units that were removed are thus censored data. The first ten observations are shown in Table 8.1.

Table 8.1. Reliable Dataset from MINITAB

Failure Time (80 Degrees)	Censoring Indicator (80 Degrees)	Failure Time (100 Degrees)	Censoring Indicator (100 Degrees)
50	1	101	0
60	1	11	1
53	1	48	1
40	1	32	1
51	1	36	1
99	0	22	1
35	1	72	1
55	1	69	1
74	1	35	1
101	0	29	1

Taken from MINITAB

Steps 1-3 of the Kaplan Meier calculation above are illustrated using the first 10 observations of the 80 degree times. Each step corresponds to a column in Table 8.2.

Note that the survival function is not calculated at the censored event times of 99 and 101. As an example, one may say that the probability of a winding not failing before 53 days is 60%, or that the probability of a winding lasting almost 74 days is 30%. The estimate of survival over a given time interval applies until just before the subsequent interval.

Table 8.2. Kaplan Meier Calculation Steps

Step1	Step2	Step3
80 Degree Event Times (in ascending order)	$\dfrac{n_j - d_j}{n_j}$	$S(t_i) = \displaystyle\prod_{j=1}^{k} \dfrac{n_j - d_j}{n_j}$
35	(10-1)/10	0.9
40	(9-1)/9	(0.888)(0.9)=0.799
50	(8-1)/8	(0.875)(0.799)=0.699
51	(7-1)/7	(0.857)(0.699)=0.599
53	(6-1)/6	(0.833)(0.599)=0.499
55	(5-1)/5	(0.800)(0.499)=0.399
60	(4-1)/4	(0.750)(0.399)=0.300
74	(3-1)/3	(0.667)(0.300)=0.200
99	/	/
101	/	/

The Kaplan Meier estimate of the survival curve for the above data may also be calculated in MINITAB (after the engine winding failure times data set 'reliable.mtw' is opened) using the following menu commands:

Under Stat, Reliability/Survival, Distribution Analysis (Right Censoring), Nonparametric Dist Analysis
- double click on C1 Temp80 to add to the **Variables** field
- under **Censor**, double click on C2 Cens80 to add it to the **Censoring Options** field, enter 0 in the **Censoring Value** field, click OK, click OK

This results in summary survival information (such as the number of censored and uncensored observations, the median survival time, and the mean survival time) and the Kaplan Meier survival estimates (in bold below) being printed in the session window.

MINITAB Output

```
Results for: reliable.MTW
Distribution Analysis: Temp80
Variable:   Temp80
Censoring Information          Count
Uncensored value                 8
Right censored value             2
Censoring value:  Cens80 = 0

Nonparametric Estimates

   Standard       95.0% Normal CI
   Mean(MTTF)      Error      Lower       Upper
    56.6000       4.4752    47.8288     65.3712
```

```
Median =      53.0000
IQR =         24.0000  Q1 =    50.0000  Q3 =    74.0000

Kaplan-Meier Estimates
            Number    Number    Survival   Standard    95.0% Normal CI
Time        at Risk   Failed    Probability  Error     Lower     Upper
35.0000     10        1         0.9000     0.0949     0.7141    1.0000
40.0000     9         1         0.8000     0.1265     0.5521    1.0000
50.0000     8         1         0.7000     0.1449     0.4160    0.9840
51.0000     7         1         0.6000     0.1549     0.2964    0.9036
53.0000     6         1         0.5000     0.1581     0.1901    0.8099
55.0000     5         1         0.4000     0.1549     0.0964    0.7036
60.0000     4         1         0.3000     0.1449     0.0160    0.5840
74.0000     3         1         0.2000     0.1265     0.0000    0.4479
```

Taken from MINITAB

The reader should explore the other analysis options available in MINITAB once a Nonparametric Distribution Analysis has been selected. One can select various graphical options for plotting the survival curve, such as adding 95% confidence interval points around the curve. The plot of the survival function with 95% confidence limits that MINITAB produces appears in Figure 8.1.

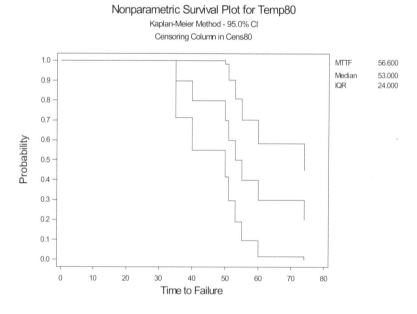

Figure 8.1. Kaplan Meier survival plot.

Various storage options are available, so that the estimates of the survival function, corresponding times, and confidence interval information can be saved to columns of data for further manipulation. These options are seen by clicking Storage once the Nonparametric Distribution Analysis-Right Censored has been selected from the MINITAB menu commands.

Two-Sample Topics

Proportional Hazards

One important concept in survival analysis is that of proportional hazards. The hazard function is defined as the instantaneous risk of failure. If two hazard functions are proportional, this implies one is a constant multiple of the other. This may be expressed mathematically as

$$h_1 = ch_2$$

for c a constant. Since the hazard function is mathematically related to the survival function, it may be shown that proportional hazards between two groups implies the survival functions of those two groups do not cross:

$$S_1 = S_2^c$$

where c is the same constant. One can verify the assumption of proportional hazards between two groups by plotting (on the same graph) the Kaplan Meier estimate of survival for each group. If the lines don't cross, the proportional hazards assumption is accurate. The importance of the proportional hazards assumption is elaborated on subsequently, but basically it is required for one of the most popular types of survival data analysis called the Cox Proportional Hazards Model.

Log Rank Test

The Kaplan Meier calculation above provides a descriptive estimate of the survival function over the course of time. However, it may be of interest to compare two groups to try and determine whether or not the groups differ with respect to survival over time. The Log Rank test is used in such a situation. It examines the departure between the observed and expected number of failures (the expected number of failures is calculated under the assumption of no differences between groups). The Log Rank test is also called the Mantel-Cox and Peto-Mantel-Haenszel test, although this text will refer to it as the Log Rank test.

The Log Rank Test Statistic is calculated using the following formulas:

$$LR = \frac{U_L^{\,2}}{V_L} \sim \chi^2_{(1)}$$

where

$$U_L = \sum_{j=1}^{r}(d_{1j} - e_{1j})$$

$$e_{1j} = \frac{n_{1j}d_j}{n_j}$$

and

$$V_L = \sum_{j=1}^{r} v_{1j}$$

$$= \sum_{j=1}^{r}\left(\frac{n_{1j}n_{2j}d_j(n_j - d_j)}{n_j^{\,2}(n_j - 1)}\right).$$

In the above formulas,

e_{1j} = the expected number of failures in the first group at the jth failure time
d_{1j} = the observed number of failures in the first group at the jth failure time
n_{1j} = the number of units at risk of failure in the first group at the jth failure time.

Similar notation holds for the second group.

It is evident from the above formulas that there can be more than one failure at a given failure time. For example, more than one engine could fail at a given time in the engine winding time data. This is a situation that is referred to as tied failure time data. There are several ways to statistically deal with tied failure times in survival data analysis. The interested reader is referred to more advanced material on this topic for an in depth discussion of various methods for dealing with tied survival data [4].

Example Table 8.3 outlines the calculations for the Log Rank test for differences in survival between two groups for the winding time data. We are interested in examining differences in failure time between the 80 degree and 100 degree groups. The calculation is based on the first seven failure times in each group.

Calculation Steps:
1. Rank in ascending order, the failure times for both groups combined. Note that only failure times are listed, and not times of engines that were censored (so that 99 and 101 are not listed).
2. Place a 1 in the second column if the failure time from column 1 corresponded to a failure in the 80 degree group, a 0 otherwise.

Table 8.3.   Log Rank Test Calculation Steps

Calculation Steps:								
1	**2**	**3**	**4**	**5**	**6**	**7**	**8**	**9**
Fail Time	d_{1j}	n_{1j}	d_{2j}	N_{2j}	d_j	n_j	$e_{1j} = \dfrac{n_{1j}d_j}{n_j}$	$v_{1j} = \dfrac{n_{1j}n_{2j}d_j(n_j - d_j)}{n_j^2(n_j - 1)}$
11	0	7	1	7	1	14	(7)(1)/(14) =0.500	(7)(7)(1)(13)/(14)(14)(13) =0.250
22	0	7	1	6	1	13	(7)(1)/(13) =0.538	(7)(6)(1)(12)/(13)(13)(12) =0.249
32	0	7	1	5	1	12	(7)(1)/(12) =0.583	(7)(5)(1)(11)/(12)(12)(11) =0.243
35	1	7	0	4	1	11	(7)(1)/(11) =0.636	(7)(4)(1)(10)/(11)(11)(10) =0.231
36	0	6	1	4	1	10	(6)(1)/(10) =0.6	(6)(4)(1)(9)/(10)(10)(9)=0 .240
40	1	6	0	3	1	9	(6)(1)/(9) =0.667	(6)(3)(1)(8)/(9)(9)(8)=0.2 22
48	0	5	1	3	1	8	(5)(1)/(8) =0.625	(5)(3)(1)(7)/(8)(8)(7)=0.2 34
50	1	5	0	2	1	7	(5)(1)/(7) =0.714	(5)(2)(1)(6)/(7)(7)(6)=0.2 04
51	1	4	0	2	1	6	(4)(1)/(6) =0.667	(4)(2)(1)(5)/(6)(6)(5)=0.2 22
53	1	3	0	2	1	5	(3)(1)/(5) =0.600	(3)(2)(1)(4)/(5)(5)(4)=0.2 40
60	1	2	0	2	1	4	(2)(1)/(4) =0.500	(2)(2)(1)(3)/(4)(4)(3)=0.2 50
72	0	1	1	2	1	3	(1)(1)/(3) =0.333	(1)(2)(1)(2)/(3)(3)(2)=0.2 22
Total	6						6.963	2.807

3. Indicate how many engines are still at risk of failure in the 80 degree group at each time.

4. Place a 1 in the fourth column if the failure time from column 1 corresponded to a failure in the 100 degree group, a 0 otherwise.
5. Indicate how many engines are still at risk of failure in the 100 degree group at each time.
6. Indicate the total number of engine failures occurring at each time.
7. Indicate the total number of engines still at risk of failure at each time.
8. Calculate

$$e_{1j} = \frac{n_{1j}d_j}{n_j}$$

at each time.
9. Calculate

$$V_{1j} = \frac{n_{1j}n_{2j}d_j(n_j - d_j)}{n_j^2(n_j - 1)}$$

at each time.
10. Sum the columns corresponding to steps 2, 8, and 9. This is the total row at the bottom of the table.
11. Calculate

$$LR = \frac{U_L^2}{V_L}$$

$$= \frac{(6 - 6.963)^2}{2.807}$$

$$= 0.33$$

as the final value of the Log Rank statistic. Note that since the value of the statistic is close to zero, it is unlikely that there is a significant difference between the 80 degree engines and the 100 degree engines. Below the MINITAB output yields a test statistic of 0.33, which when compared to the Chi Square 0.05 (95% confidence) table value of 3.84 is not significant.

Extensions of the above Log Rank formula (which tests for differences in suvival between two groups) exist for more than two groups as well as for examining differences between groups in the presence of other variables (called the stratified Log Rank test).

MINITAB Example

To compute the Log Rank statistic for this data in MINITAB, select the following menu commands:

Under Stat, Reliability/Survival, Distribution Analysis (Right Censoring), Nonparametric Dist Analysis

- double click c1 Temp80 and c3 Temp100 to add them to the **Variables** field
- under **Censor**, double click c2 Cens80 and c4 Cens100 to add them to the **Censoring Options** field, enter 0 in the **Censoring Value** field, click OK
- under **Results**, select **In Addition, Hazard, Density Estimates and Log Rank and Wilcoxon Statistics**, click OK, click OK

This produces separate distribution analyses for the 80 degree windings and the 100 degree windings. Each distribution analysis includes the proportion of censored data, the median censoring time, the mean censoring time with 95% confidence limits and estimates of the survival function with 95% confidence limits. The output 'Distribution Analysis, Temp80, Temp100' gives the value of the Log Rank test statistic as 0.3310 (shown in bold below), confirming there is no significant difference between the 80 degree and 100 degree windings with respect to engine failure times.

MINITAB Output

```
Distribution Analysis: Temp80, Temp100
Comparison of Survival Curves
Log-Rank Statistic
Variable            1              2
                -0.9641       0.9641
Variance/Covariance of Log-Rank Statistic
Variable            1              2
  1             0.3561       -2.8081
  2            -2.8081        2.8081

Wilcoxon Statistic
Variable            1              2
               -18.0000       18.0000
Variance/Covariance of Wilcoxon Statistic
Variable            1              2
  1             0.004149    -241.0000
  2          -241.0000       241.0000

Test Statistics
Method      Chi-Square     DF     P-Value
Log-Rank       0.3310       1      0.5651
Wilcoxon       1.3444       1      0.2463
```

Taken from MINITAB

The interested reader is referred to [4] for details on the Log Rank test for multiple group survival comparisons and the stratified Log Rank test for comparisons accounting for subgroups. Both the Log Rank test and the previously discussed Kaplan Meier estimate of survival are not based on any distributional assumptions. As a next step, the assessment and application of distributional assumptions are discussed.

Distribution-Based Survival Analyses

As with most statistical techniques, application of a distribution-based or parametric analysis results in more precise test statistics. That being said, the distribution selected must be appropriate for the data. MINITAB offers a useful feature which helps identify which distribution is appropriate for a given set of failure time data. Goodness of fit for a particular distribution may be tested, the chosen distribution can be applied to the data and test statistics calculated. This process will be applied to the previously discussed MINITAB dataset 'reliable.mtw'.

Commonly used parametric failure time distributions include the Exponential, the Weibull, the Lognormal and the Normal. These distributions are selected due to features which coincide well with one of the main properties of failure time data – failure times must always be positive. See [4] for more information on properties of failure time distributions. Once a distribution has been identified as being appropriate for a set of data, the general procedure is to estimate its parameters. This is generally done using one of two methods – maximum likelihood estimation or least squares estimation. Least squares estimation was discussed in Chapter 7. The general technique in least squares estimation is to minimize the sum of the squares of the residuals. Maximum likelihood estimation determines parameter estimates iteratively so that the probability of the given data is largest. The final step in the case of one sample of data is to compare the parameter estimate with a hypothesized value or in the case of two samples to compare the parameter estimates to each other to determine whether the two samples are significantly different with respect to failure time.

MINITAB Example

Example   The following MINITAB menu commands produce a distribution assessment of the engine failure times for the 80 and 100 degree temperatures. All data is used in the analysis.

Under Stat, Reliability/Survival, Distribution Analysis (Right Censoring), Distribution ID Plot
- double click C1 Temp80 and C3 Temp100 to add them to the **Variables** field

- under **Censor**, double click C2 Cens80 and C4 Cens100 to add them to the **Censoring Options** field, enter 0 in the **Censoring Value** field, click OK, click OK

This yields the following MINITAB (truncated) and graphical output in Figure 8.2.

MINITAB Output

```
Distribution ID Plot

Variable:  Temp80

Goodness of Fit

Distribution         Anderson-Darling (adj)
Weibull              67.64
Lognormal base e     67.22
Exponential          70.33
Normal               67.73

Variable:  Temp100

Goodness of Fit

Distribution         Anderson-Darling (adj)
Weibull              16.60
Lognormal base e     16.50
Exponential          18.19
Normal               17.03
```

Taken from MINITAB

The output suggests that the log normal distribution fits the data best due to the observed points falling closest to the straight line. This is reflected by the fact that the Anderson-Darling statistic is smallest for the log normal distribution.

The next step is to fit the log normal distribution to the data and test for differences between the 80 and 100 degree temperatures.

The following MINITAB menu commands fit the log normal distribution to the data.

Under Stat, Reliability/Survival, Distribution Analysis (Right Censoring), Parametric Dist. Analysis

- double click on C1 Temp80 and C3 Temp100 to add them to the **Variables** field
- under **Censor**, double click C2 Cens80 and C4 Cens100 to add them to the **Censoring Options** field, enter 0 in the **Censoring Value** field, click OK
- under **Test, Equality of Parameters**, select Test for Equal Shape and Test for Equal Scale, click OK
- under **Assumed Distribution**, select Log normal, click OK

The MINITAB output for the log normal distribution fit follows, with graphical output in Figure 8.3.

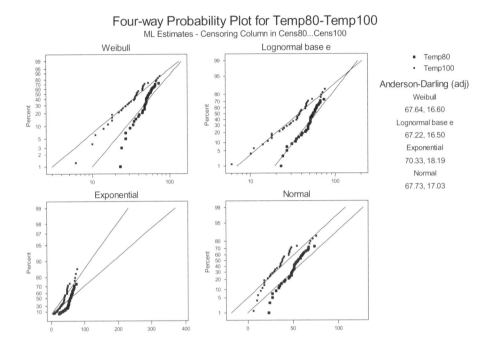

Figure 8.2.   Survival distribution identification.

The Chi Square values shown in the MINITAB output are all greater than the critical Chi Square value of 3.84 (at 95% certainty), indicating there are differences between the 80 and 100 degree temperatures. The confidence intervals shown are for the ratio of the scale parameters for the two temperatures and for the difference between the location parameters. If the two temperatures produce similar engine failure times, it is expected that the confidence interval around the ratio of the scale parameters should include 1 and the confidence interval around the difference in the location parameters should include 0. This is not the case, indicating different scale and location parameters for the two temperatures. Such a conclusion is further supported by Figure 8.3, for which the estimates of the survivor function and confidence intervals only overalp at later failure times. It appears that compared to the 100 degree temperature, the 80 degree temperature produces engines that fail later. It is interesting to note that the Log-Rank test (which had no distributional assumptions) based on the first seven data points, produced a test statistic which indicated there were no significant differences between the temperatures. Using a parametric distribution (in this case the log normal) to analyze the entire dataset has resulted in detection of differences between the temperatures.

MINITAB Output

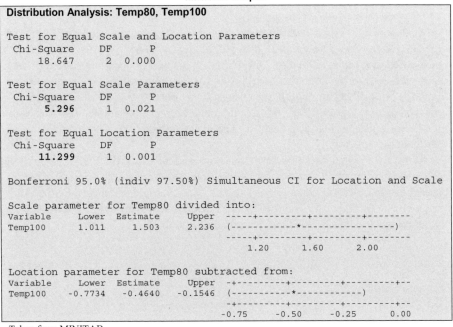

```
Distribution Analysis: Temp80, Temp100

Test for Equal Scale and Location Parameters
  Chi-Square    DF      P
    18.647       2   0.000

Test for Equal Scale Parameters
  Chi-Square    DF      P
     5.296       1   0.021

Test for Equal Location Parameters
  Chi-Square    DF      P
    11.299       1   0.001

Bonferroni 95.0% (indiv 97.50%) Simultaneous CI for Location and Scale

Scale parameter for Temp80 divided into:
Variable    Lower   Estimate    Upper   -----+---------+---------+--------
Temp100     1.011    1.503      2.236    (------------*----------------)
                                         -----+---------+---------+--------
                                             1.20      1.60      2.00

Location parameter for Temp80 subtracted from:
Variable     Lower    Estimate     Upper   -+---------+---------+---------+--
Temp100    -0.7734    -0.4640    -0.1546    (----------*------------)
                                            -+---------+---------+---------+--
                                          -0.75     -0.50     -0.25      0.00
```

Taken from MINITAB

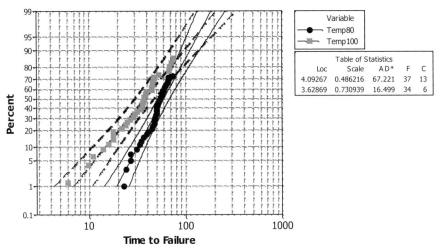

Probability Plot for Temp80, Temp100
Lognormal - 95% CI
Censoring Column in Cens80, Cens100 - ML Estimates

Figure 8.3.   Comparing log normal models for reliable dataset.

Summary

Suvival data analysis has rapidly evolved over the past twenty years, becoming one of the most widely researched areas in statistics. The previous material has introduced concepts pivotal to survival data analysis, such as censored data. Statistical methods were discussed for estimation and summary of the survival function as well as for non-parametric and parametric testing of differences in survival between groups. This information only scratches the surface in terms of the possibilities for survival data analysis. The interested reader is referred to [4] or [5] for more information.

As a final note, the popular Cox Proportional Hazards Model should be mentioned as a method for comparing two groups with respect to survival. The Cox model has the pleasing feature that it is semi-parametric. This means that the only requirement for the data is proportional hazards. No distributional form for the failure times is conjectured. The Cox model is written

$$h_1(t) = e^{\beta x_i} h_0(t) \, .$$

If one is interested in comparing failure times of experimental units manufactured using two different processes (standard=0 and new=1), the verbal interpretation is that the hazard of failure for a given unit is a constant (which depends on whether the given unit was produced using the standard or new process) multiplied by the hazard of failure for the standard unit. While MINITAB Release 14 does not currently offer menu commands for Cox PH modeling, macros may be developed within MINITAB for this type of analysis. MINITAB macros are discussed further in Chapter 9. Reference 6 provides more information concerning survival analyses available in MINITAB.

TIME SERIES DATA

Introduction

In a similar fashion to survival data, time series data are also concerned with times of events. That being said, there are fundamental differences between the two types of data. The previous section showed that survival data focuses on the time of a specific event for a given unit/person. This event may or may not occur, so that censoring plays a role in survival data analysis. We observe the event times (or last time on study for censored data) for n units, so that the data appears as

$$t_1, t_2, \ldots t_n$$

with corresponding censoring information. Time series data focuses on the recorded values for a given measurement at several points in time. This type of data can be expressed as

$$y_1, y_2, \cdots y_n$$

where y_i denotes the measured value at time i. The data is not censored in the sense that the investigator is never waiting for an event to happen which may not occur. With time series data, the investigator picks the times at which the desired measurements are taken. In survival data, the investigator records the times of a specific event. Each type of analysis poses its own set of challenges. For survival data, this is the inclusion of censored data in a meaningful manner. For time series data, this is the dependence that is encountered when the same measurement is taken repeatedly over time. Recall that one of the basic requirements for statistical analysis validity (discussed in Chapter 3) is that the data be independent. This was defined by each data point being uninfluenced by any other data point in the set. It is now quite obvious that when the same exact measurement is taken repeatedly, many types of influence may exist. For example, monthly temperature readings are highly dependent on season, so that the July temperature will be closely related to temperatures taken in other summer months. Time series data may even be cumulative, which by nature implies arithmetic dependence. Of course dependence between data can be encountered for any type of data previously discussed, including normally distributed, proportions, and survival data. For these types of data, dealing with dependence is an extension of the basic situation in which the data is independent. A basic assumption of dependence in the data underlies time series analyses. Although not discussed here, another large component of time series analyses includes forecasting, or predicting the values of future measurements based on past observations. The interested reader is referred to [7].

Data Presentation

This section discusses basic techniques for graphical presentation of time series data. The reader is referred to [7] for a more detailed statistical discussion and analyses.

Time Series Plots

One of the most simple yet informative ways of becoming familiar with data is to display it graphically. This is especially true for time series data.

MINITAB Example

Consider the MINITAB dataset 'cranksh.mtw'. A manager in an engine production facility is concerned with measuring the distance from a point on the crankshaft of the engine to the baseline position. This is done five or ten times each day on nineteen different days. The first ten observations in the dataset appear in Table 8.4:

Table 8.4.   Crankshaft Dataset from MINITAB

AtoBDist	Month	Day
-0.44025	9	28
5.90038	9	28
2.08965	9	28
0.09998	9	28
2.01594	9	28
4.83012	9	29
3.78732	9	29
4.99821	9	29
6.91169	9	29
1.93847	9	29

Taken from MINITAB

The measurements continue over month 10 as well. There are several ways to proceed when examining this time series. One could take the average of the five or ten measurements on each day, and then examine the resulting time series. This requires some manipulation of the data. Another column of data that groups samples taken on the same day is added so that each day's data can be averaged. The new variable 'Time' is shown in Table 8.5 (the last two rows of the dataset are added for clarity).

The following MINITAB menu commands take the mean of AtoBDist within each level of the Time variable, giving the mean for all nineteen timepoints:

Under Stat, Basic Statistics, Store Descriptive Statistics
- double click on C1 AtoBDist to add it to the **Variables** field
- double click on C4 Time to add it to the **By Variable** field, click OK
- under **Statistics**, select mean, click OK, click OK

Columns for Byvar1, Mean1, and N1 have been added to the spreadsheet. Data is entered for another variable, Stamp, which reflects the day and month of the average. The first five columns in the revised dataset are shown in Table 8.6.

Table 8.5.   Crankshaft Dataset Revised

AtoBDist	Month	Day	Time
-0.44025	9	28	1
5.90038	9	28	1
2.08965	9	28	1
0.09998	9	28	1
2.01594	9	28	1
4.83012	9	29	2
3.78732	9	29	2
4.99821	9	29	2
6.91169	9	29	2
1.93847	9	29	2
⋮	⋮	⋮	⋮
0.27517	10	25	19
-5.32797	10	25	19

Taken from MINITAB

Table 8.6.   Crankshaft Means by Time

Byvar1	Mean1	N1	Stamp
1	1.93314	5	Sept28
2	4.49316	5	Sept29
3	-0.72507	5	Sept30
4	3.22633	5	Oct4
5	3.88394	5	Oct5

Taken from MINITAB

The following menu commands may then be used to produce a time series plot of the average distance measured each day during the nineteen days (Figure 8.4):

Under Graph, Time Series Plot
- select **Simple**, click OK
- double click on Mean1 to add it to the **Series** field
- under **Time Scale,** select Stamp, double click C8 Stamp to add it to the **Stamp Columns** field
- under **Labels,** enter a title in the **Title** field, a footnote in the **Footnote** field, click OK, click OK

It is difficult to draw many conclusions regarding the crankshaft data. The average distance varies considerably over the nineteen day period. The averages based on ten observations appear less variable than those based on only five observations.

Smoothing

The time series plot of the average crankshaft distance indicated that the data was quite variable. One commonly used technique for addressing variability in time

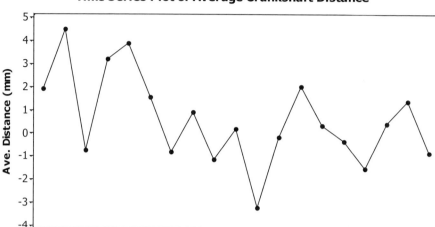

Sept 28-Oct 15 averages are based on 5 observations
Oct 18- Oct 25 averages are based on 10 observations

Figure 8.4.   Time series plot.

series is known as smoothing. Smoothed data converts each measurement into the average obtained using the measurement in question with measurements before and after:

$$y_i^* = \frac{y_{i-1} + y_i + y_{i+1}}{3}$$

Smoothing data retains major features of the data and helps to remove variability.

MINITAB Example

The following MINITAB menu commands convert the nineteen means to smoothed observations and produce a plot of the smoothed observations:

Under Stat, Time Series, Moving Average
- double click Mean1 to add it to the **Variable** field
- enter 3 in the **MA length** field
- check Center the Moving Averages
- under **Storage**, select Moving Averages, click OK

- under **Graphics**, select Plot Smoothed vs Actual, click OK
- under **Options**, enter 'Smoothed Average Crankshaft Distance' in the **Title** field, click OK

This produces the column Aver1 (the smoothed values) and the following output:

Crankshaft Measurements–Smoothed Means

Aver1
*
1.90041
2.33147
:
0.31969
*

Taken from MINITAB

Note that the first and last smoothed values can't be calculated due to the average requiring the term before and after to be present for the smoothing, i.e., the first value is missing the time before and the last value is missing the time after. This was specified by the 3 in the MA length field in MINITAB. The reader should verify the calculation of the average for each of the entries.

MINITAB Output

```
Moving Average

Data        Mean1
Length      19.0000
Nmissing    0
Moving Average
Length: 3

Accuracy Measures
MAPE: 204.162
MAD:    1.581
MSD:    3.435
```

Taken from MINITAB

MINITAB also produces a graph of the smoothed data, superimposed on the original data (Figure 8.5).

The smoothed data indicate that the average crankshaft distance between points A and B is decreasing over time. Note that MINITAB offers an option for exponential smoothing in the time series menu commands. This differs from the arithmetic smoothing discussed above.

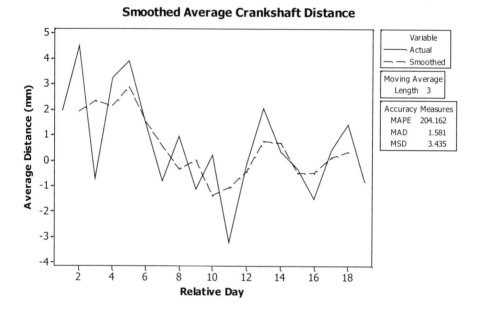

Figure 8.5.   Smoothed time series plot.

Moving Averages

Smoothing is actually a special case of what is referred to as a moving average. A moving average defines the observation at each time point y_t^{MA} as an average of a specified number of surrounding observations y_{t+q}, with specified weights w_q :

$$y_t^{MA} = \sum_{q=-r}^{r} w_q y_{t+q} \quad t = r+1,\dots n-r$$

Although the formula may appear complicated, it is simpler to interpret this as making each point an average of the entire series, with different weights given to various terms in the average. When calculating a moving average, one must specify the length or order of the average, which is essentially the number of terms to use in calculating the average. This is always an odd number since the moving average value for a given term y_t will always include that term itself and then select pairs on each side of the term

$$(y_{t-1}, y_{t+1}), (y_{t-2}, y_{t+2}), \ldots (y_{t-k}, y_{t+k}).$$

A moving average of length five by definition leaves the first four observations undefined. The higher the length or order of the average, the more smoothing that occurs. The smoothing discussed earlier was actually a moving average of length 3.

It is informative to compare the moving average of length 3 calculated in the smoothing example to one of length 5. The moving average of length 5 is calculated for the crankshaft data using the same menu commands as for length 3.

MINITAB Example

The following menu commands produce the overlayed moving average graph for the crankshaft data:

Under Graph, Time Series Plot
- select **Multiple**, click OK
- double click on c9 MA(3) and c10 MA(5) to add them to the **Series** field
- under **Time Scale**, select Stamp, double click C8 Stamp to add it to the **Stamp Columns** field, click OK
- under **Labels,** enter a title in the **Title** field, a footnote in the **Footnote** field, click OK, click OK

The graph of the averaged data appears in Figure 8.6.

It is clear that as the length of the moving average gets longer, variability is removed from the data.

Summary

The previous section introduced basic time series presentation techniques, such as the time series plot and moving averages. MINITAB menu commands for graphical displays were provided. Many other more sophisticated time series analyses are discussed further in [7].

MINITAB offers the user a wide variety of options for analysis of time series data.

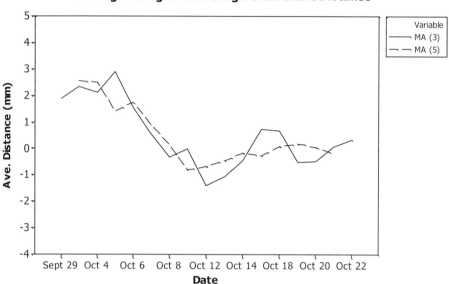

Figure 8.6. Moving averages of crankshaft dataset.

EXERCISES

8-1. A metrologist samples 25 materials and finds 22 are acceptable. Generate a two-sided 90% confidence interval around the proportion of acceptable materials. After further consideration, the metrologist realizes he is only concerned with the lower bound of the confidence interval about the proportion of acceptable materials. Generate a one-sided lower 90% confidence interval around the proportion of accpetable materials. Explain the difference between the two confidence intervals. Repeat these calculations using MINITAB software.

8-2. The dataset 'billiard.mtw' in MINITAB gives the elasticity of billiard balls produced in different batches under 3 conditions (no additive, additive 1, additive 2). It has been determined that billiard balls with elasticity less than 42 are unacceptable. In MINITAB, convert the elasticity measurement to a new variable indicating whether that ball was acceptable or not. Use MINITAB to calculate a 95% one-sided lower confidence interval around the proportion of acceptable billiards. Use MINITAB to calculate a 99% confidence interval around the difference in proportion of acceptable billiards between those with no additive (0) and those with either additive 1 or 2. Hint: Create another variable in MINITAB to indicate whether each billiard was produced using no additive or any type of additive. Disregard batch for all calculations.

8-3.   The table below presents time of preservative degradation for two different preservatives.

Preservative Degradation Time

Degradation Time Preservative 1 (msec)	Censoring Indicator Preservative 1	Degradation Time Preservative 2 (msec)	Censoring Indicator (Preservative 2)
540	1	665	0
690	1	785	0
720	1	598	1
495	1	512	1
615	1	516	1
555	0	584	1
810	0	702	1

Calculate the Kaplan Meier estimate of degradation time. Plot the estimates on the same graph. Does the assumption of proportional hazards appear justifiable? Explain. Use the Log Rank test to examine differences in degradation between the two preservatives. Repeat the above calculations using MINITAB. Using MINITAB, fit the best parametric distribution available to the data. Justify your choice. Based on this distribution, are there significant differences between the preservatives?

8-4.   Using the MINITAB dataset 'employ.mtw', construct an overlayed time series plot showing the number of employees for each industry. Construct an overlayed plot of the smoothed data. What differences are apparent between the original and smoothed data within each industry?

8-5.   One can find many valuable datasets on the internet. The websites 'www.amstat.org/publications/jse' and 'www.blackwellpublishing.com/rss' provide several time series datasets. Select one of them for analysis. Construct time series and smoothed plots of the data. Examine differences between moving averages of various lengths.

REFERENCES

[1]   Harnett, D. L., Murphy, J. L. *Statistical Analysis for Business and Economics* (New York: Addison-Wesley, 1993).

[2]   Natrella, M.G. "Experimental Statistics," NBS Handbook 91, (Gaithersburg, MD: National Institute of Standards and Technology 1963, p 3-40).

[3]   Fleiss, J. *Statistical Methods for Rates and Proportions Second Edition* (New York: John Wiley and Sons, 1981).

[4]   Collett, D. Modeling Survival Data in Medical Research (London: Chapman and Hall, 1994).

[5]   Shoukri, M. M., Pause, C. A. *Statistical Methods for Health Sciences* (Boca Raton, FL: CRC Press, 1998).

[6]   *MINITAB User's Guide 2: Data Analysis and Quality Tools,* Release 14 for Windows (Minitab Inc. 2003, p 9-7)

[7]   Diggle, P. J. *Time Series: A Biostatistical Introduction* (Oxford: Clarendon Press, 1990)

CHAPTER 9

Selected Topics

The statistical techniques discussed in the previous chapters are the ones that the author has found to be most useful for planning and evaluating measurement data. There are, of course, many other techniques that are very useful in special situations and may even be the techniques most frequently used in some applications. Some of these are described briefly in the present chapter.

BASIC PROBABILITY CONCEPTS

The theory of probability is not only basic to statistics but it is involved in many day-to-day situations. One frequently hears such statements as the chance of something occurring is less than "one in a million", or there is a 30% chance for something, say rain, to occur. Such statements are often based on little more than guesses and need to be so considered. However, the principles upon which true probability statements can be made are well established and the ones that are most useful will be described below.

- If an event can occur in a ways and not occur in b ways, and all ways are equally probable, the probability of occurrence is

$$\text{Probability of occurrence of a} = \Pr(a)$$
$$= \frac{a}{a + b}$$

- The probability of not occurring is

$$\text{Probability of a not occurring} = \Pr(\bar{a})$$
$$= \frac{b}{a+b}$$

- The probability of occurrence of an event is a proper fraction varying between 0 (impossible) and 1 (certainty).
- If an event can happen in two or more independent ways whose respective probabilities are known, the probability of the occurrence of the event is equal to the sum of the probabilities of its happening in the several ways.

Consider a box containing R red balls, W white balls, and B blue balls. The probability, P, of randomly selecting a ball of a specified color is

$$P_R = \frac{R}{R+W+B}, \quad P_W = \frac{W}{R+W+B}, \quad \text{and} \quad P_B = \frac{B}{R+W+B}$$

respectively. The probability of selecting either a red or a white ball is

$$P_{R \text{ or } w} = P_R + P_W = \frac{R+W}{R+W+B}$$

- If two or more independent events can occur simultaneously and the probabilities of the separate events are known, the probability of the simultaneous occurrence will be the product of the respective probabilities.

In the case of the example given above, the probability of selecting two red balls in a drawing is

$$P_R \times P_R = \frac{R^2}{(R+W+B)^2}$$

provided that the selected ball is put back in the bowl after each withdrawal.

- The expected number of occurrences of a specific event is the product of the probability and the number of attempts. The word *expected* is a key word.

There is no guarantee that the expected will happen, except in the case of many trials. Thus, there is a probability, P = 0.5, that the flip of a coin will result in a "heads" but there is no guarantee that 2 flips will result in a "heads". You can believe from what has been said above that there is a small chance, actually 0.0078125 or 1 in 128 [i.e., $(1/2)^7$] that a heads will not appear in 7 successive flips.

It should always be remembered that the probability of nonoccurrence is equal to one minus the probability of occurrence, so the probability that at least one heads will occur during 7 successive flips is 1 − .0078125 or 0.9921875 which is equivalent to 127 in 128.

Statistical probability is based on consideration of areas under a distribution curve. Thus, the area defined by plus and minus one sigma from the mean in a normal distribution represents 68.7% of the total area under the curve and the probability of occurrence of values differing from the mean by no more than that amount is 0.687 or 68.7%. There is a probability of 1 − 0.687 or 0.317 (almost 1 in 3) that the result of a single measurement will be outside of these limits.

Statistical probabilities depend on knowing what kind of distribution is present and the availability of sufficient data to define it. We have seen already that the important parameters of a normal distribution are the mean, μ, and the standard deviation, σ. For small samples, these are poorly estimated (and sigma is often underestimated) hence confidence limits based on the estimates $\overline{X}$ and s need to be larger than those for which μ and σ are known.

MEASURES OF LOCATION

The subject of selection of a central value as a descriptor of a data set has been discussed in Chapter 3. The following discussion provides further insight into this matter.

Mean, Median, and Midrange

The idea that the average value is more reliable than the individual values of a set of repetitive measurements hardly needs defense, yet this was not recognized until postulated and advocated by Thomas Simpson in 1755. The discussion of Eisenhart [1] concerning this makes interesting reading. The selection of a central value as a descriptor of a population characteristic is considered in Chapter 3 where the arithmetic mean and geometric mean are discussed. The subject of weighting when combining means was considered in Chapter 6. Individual values in a data set may be weighted differently if there is reason to do so. Averaging of such values results in the computation of a weighted mean. Ordinarily, all values in a set are assigned either a weight of 1 or 0. A zero weight is assigned to a value that is to be disregarded, such as an outlier, for example. All values receiving a weight of 1 are given equal consideration when calculating a simple arithmetic mean. Assignment of a weight greater than 1 means that such a value is believed to be of greater importance when calculating the average.

The median is the middle value. Its use minimizes the influence of unbalance due to extreme values, either large or small, in the calculation of the mean. The median and the mean are identical for a normal distribution but can differ considerably when evaluated using small samples (small number of data points). The midrange is the average of the largest and smallest data points of a set. It provides ease of calculation but is sensitive to extreme values in the tails of a distribution.

Measures of location is a large subject in itself. It is discussed in considerable depth by Crow and Siddiqui [2]. One may choose the measure of location, m, such that:

$$\Sigma(X_i - m)^2 = \text{minimum}; \quad \text{unweighted mean is the best choice}$$
$$\Sigma(X_i - m) = \text{minimum}; \quad \text{median is the best choice}$$
$$\max|X_i - m| = \text{minimum}; \quad \text{midrange is the best choice}$$

The unweighted mean is considered to be the best estimator in the case of a normal distribution, the median is best if the distribution is double exponential, and the midrange is best if it is rectangular [3].

Trimmed Means

The adverse affect of unequal tails on the calculated value of a mean may be eliminated by outlier elimination or more simply by deliberately discarding a percentage of data on each end. Thus, a 25% trimmed mean results from trimming 25% of the data from each tail and basing the mean on the remaining 50% of the original data. This procedure is sometimes called "Winsorizing". The mean and median of such a trimmed data set usually agree quite well. The practice of using a trimmed mean is not suitable when small data sets are involved. Twelve points is believed by some to constitute a minimum data set when employing a trimmed mean.

The trimmed body of data described above provides a quick way to estimate the standard deviation by what is sometimes called "polishing". Since 50% of the data is expected to lie within $\pm 2/3$ s,

$$s \approx 1.5 \text{ times the half width of the 25\% trimmed data set.}$$

Average Deviation

The average deviation is computed as

$$\text{Avg. Dev.} = \Sigma|X_i - \overline{X}| / n$$

Note that the absolute values are summed. If signs are taken into account, the sum is, of course, zero! The average deviation was used extensively by physical scientist for a long time. It is easy to compute and is less affected by outliers than the standard deviation. For large samples of normally distributed data,

$$\text{Average Deviation} \approx 0.8\,\sigma$$

Table 9.1. Ratio of Average Deviation to Sigma for Small Samples

Degrees of Freedom	Range for Various Levels of Confidence		
	99%	95%	90%
10	.6675—.9359	.7153—.9073	.7409—.8899
15	.6829—.9137	.7236—.8884	.7452—.8733
20	.6950—.9001	.7304—.8768	.7495—.8631
25	.7040—.8901	.7360—.8686	.7530—.8570
30	.7110—.8827	.7404—.8625	.7559—.8511
35	.7167—.8769	.7440—.8578	.7583—.8468
40	.7216—.8722	.7470—.8540	.7604—.8436
45	.7256—.8682	.7496—.8508	.7621—.8409
50	.7291—.8648	.7518—.8481	.7636—.8385

Adapted from Reference 3.

Departures from this value are an indication of nonnormal distribution. Table 9.1 lists the range of values for the ratio, average deviation (Avg. Dev.)/σ that might be expected to be found for small samples of normally distributed data, due to variability considerations. As an example, departures from the ratio Avg. Dev./Standard Deviation for normally distributed sets of 16 points would be expected to lie within the limits 0.7236 to .8884, 95 out of 100 times. There would be a 95% confidence that if the data are normally distributed, the value found would be inside of this range. Values for 90 and 99% confidence decisions are also listed in the table.

TESTS FOR NONRANDOMNESS

It was pointed out in Chapter 3 that randomness of the data is a basic requirement for valid statistical analysis. While randomness cannot be defined quantitatively, systematic trends in data often can be identified. Simple plotting may be sufficient in many cases to identify gross departures from randomness but this is a subjective decision. Fortunately, there are several statistical tests that can be used, as well, and several of them are described below. Unfortunately, most tests become less useful when small data sets are of concern.

Runs

Runs consist of consecutive data points that exhibit a pattern of generally increasing or decreasing in value. If plotted, they may be represented by a positive or a negative slope, respectively. Runs can also consist of a group of points that stay on one side of a residual plot or of the central line of a control chart or a sequential data plot. Since limited runs can appear in random data, one has to distinguish real runs from those which occur due to chance fluctuations. One can look for runs in data sets that are plotted sequentially, in the distribution of residuals from a calculated mean or of experimental points from a fitted line, and points plotted on a control chart.

Runs in a Data Set

List the data sequentially and calculate the average value of the data set. Place a + sign over each value that is larger than $\overline{X}$ and a − sign over each smaller value. Assign a + sign over any value that is equal to $\overline{X}$. The number of runs is equal to the number of changes of sign + 1.

It can be shown that the expected number of runs in a random data set of n points can be calculated from the expression [3]

$$Expected\ Number\ of\ Runs = 1/3\,(2n\text{-}1)$$
$$S.D.\ of\ Number\ of\ Runs = [(16n\text{-}29)/90]^{1/2}$$

To determine whether an excessive number of runs (or too few runs) occur in a data set, calculate the ratio

$$ratio = \frac{Observed\ Runs - Expected\ Runs}{S.D.\ of\ Number\ of\ Runs}$$

In the case of n > 19, compare the ratio with the Z values in Table A.2 in Appendix A to determine the probability of its being due to chance (i.e., 1 − table value = probability of occurrence due to chance causes). In the case of a smaller number of values, use the t table, Table A.3 in Appendix A, to make the determination. (Note degrees of freedom = n − 1 for deviations from a mean and n − 2 for residuals from a plotted line). In general, a ratio greater than 2 indicates low probability and one greater than 3 indicates a very low probability of occurrence due to chance, and thus more or less runs (depending on the sign of the difference) than would be expected by chance alone, i.e., something nonrandom would seem to be operating.

Short-term drifts would tend to yield less runs than expected. Cyclic data would yield either larger or smaller numbers of runs, depending on the period of the cycles.

Runs in Residuals from a Fitted Line

In the case of deviations from a fitted line, use the following procedure. Plot the data and indicate a value above the line by a + sign and a value below the line by a − sign. Arbitrarily indicate a value on the line by a + sign. Let

$$M_1 = number \text{ of } + deviations$$
$$M_2 = \text{number of} - deviations$$

Each change in the sequence of signs is a run. The total number of runs is equal to the number of changes in sign plus one.

It can be shown [3] that

$$\text{Expected number of runs} = 1 + \frac{2M_1 M_2}{n}$$
$$\text{S.D. of Runs} = [\frac{2M_1 M_2 (2M_1 M_2 - n)}{n^2 (n-1)}]^{1/2}$$

Calculate the ratio

$$\text{ratio} = \frac{|Observed \text{ runs} - \text{Expected runs}|}{\text{S.D. of Number of Runs}}$$

and interpret the ratio as described in the previous section.

Trends/Slopes

The significance or apparent trends with respect to time or sequence of observation can be judged for entire data sets or portions of a series of data points in the following manner. The data points of concern are fitted by a straight line using the method of least squares. The slope and the standard deviation of the slope are recorded. Obviously, a slope significantly different from zero indicates a trend. A slope of zero is rarely found but the statistical significance of that found can be decided using the following ratio

$$t = \frac{\text{Slope} - 0.000}{\text{S.D. of Slope}}$$

and referring to a t table such as Table A.3 in Appendix A. The number of degrees of freedom used in selecting the value for t is (n − 2), where n is the number of data points. The sign of the ratio is important, only in that it indicates a positive or a negative slope (trend of the data).

As an example of the above, a set of data consisting of 20 points was fitted as described and a slope of −2.50 was obtained with S.D. of 1.45 Accordingly, t = 1.72. For 18 degrees of freedom (20 − 2), the closest value for t in the table is $t_{.95}$, the subscript .95 denoting the value 1 − $\alpha/2$, since a two-sided probability test is involved. Hence $\alpha/2 = 0.05$ and $\alpha = 0.10$. In other words, the probability is 0.10 or 10% that the observed slope could result from chance causes.

It should be remembered that a wild data point (outlier) can exert a large influence (unmerited) on the estimated slope obtained by a least squares fit, especially if a small number of points are being fitted. The author recommends that a preliminary graphical fit should be made before any least squares fit and any point deviating by more than 4 average deviations from the line be considered as an outlier (huge error concept) and excluded from the set that is fitted by least squares.

Mean Square of Successive Differences

The mean square of successive differences (MSSD) may be used to test for the presence of correlation between adjacent observed values which would be a violation of randomness. The calculation is made as follows:

$$MSSD = \sum_{i=1}^{n} (X_{i+1} - X_i)^2 / (n - 1)$$

It has been shown that the ratio

$$MSSD/\text{Variance} \rightarrow 2 \text{ as } n \rightarrow \infty$$

for a normal distribution. If successive differences are small in comparison with the spread of the data, then the ratio will decrease.

Example   Consider the data set

$$11, 13, 12, 14, 15, 17, 17, 18$$

The sequential differences are 2, −1, 2, 1, 2, 0, 1

$$MSSD = 2.143$$

The variance of the data set is 5.871; hence,

$$MSSD/Variance = 0.3650$$

Table 9.2. Critical Values for the Ratio MSSD/Variance

	Probability Level		
N	0.1%	1.0%	5.0%
4	0.5898	0.6256	0.7805
5	.4161	.5379	.8204
6	.3634	.5615	.8902
7	.3695	.6140	.9359
8	.4036	.6628	.9825
9	.4420	.7088	1.0244
10	.4816	.7518	1.0623
20	.7852	1.0406	1.2996
40	1.0850	1.2934	1.4921
60	1.2349	1.4144	1.5814
∞	2.0000	2.0000	2.0000

Excerpted from Reference 3. An extensive table will be found in Reference 4.

This is much smaller than 2 but the data set is small and one would expect that a smaller number for the ratio might happen just by chance in such cases.

Table 9.2 (included for illustrative purposes) contains the smallest values that would be expected to occur by chance alone in a small sample from a normally distributed population, at three probability levels. Reference 4 should be consulted for a more extensive table and a further discussion of this subject.

In the example above, there is a 5% probability that a number smaller than 0.9825 could occur in a random sample of size 8 (or once in 20 times, a value lower than that could occur) and a probability of 0.1% (1 in 1000) that a value smaller than

0.4036 could occur. The value in the example, 0.3650, has a probability of much less than 1% (<<1 in 100) of occurring in a random sample and confirms the existence of a run, i.e., a nonrandom event.

The MSSD analysis can be used for an entire data set or for a suspected run in a data set. Again, suspected runs in control chart data may be evaluated by the MSSD/ variance ratio.

COMPARING SEVERAL AVERAGES

In Chapter 5, the t test was described for use in deciding whether two means differed significantly. The following describes how to compare the averages of several quantities (e.g., products or measurement results) [5]. It is essentially a version of a one-way analysis of variance.

The data to be considered consists of t sets of n measured values and their means, X_1 to X_t, and their standard deviation estimates, s_t, each based on $n_t - 1$ degrees of freedom, respectively. (The symbol, t, employed here is the one used in Reference 5. It should not be confused with the Student t variate.) The only restriction is that

Table 9.3. Percentiles of the Studentized Range, $q_{.95}$

t df	2	3	4	5	6	7	8	9	10
1	18.0	27.0	32.8	37.1	40.4	43.1	45.4	47.4	49.1
2	6.1	8.3	9.8	10.9	11.7	12.4	13.0	13.5	14.0
3	4.5	5.9	6.8	7.5	8.0	8.5	8.8	9.2	9.5
4	3.9	5.0	5.8	6.3	6.7	7.0	7.4	7.6	7.8
5	3.6	4.6	5.2	5.7	6.0	6.3	6.6	6.8	7.0
6	3.5	4.3	4.9	5.3	5.6	5.9	6.1	6.3	6.5
7	3.3	4.2	4.7	5.1	5.4	5.6	5.8	6.0	6.2
8	3.26	4.04	4.53	4.89	5.17	5.40	5.60	5.77	5.92
9	3.20	3.95	4.41	4.76	5.02	5.24	5.43	5.59	5.74
10	3.15	3.88	4.33	4.65	4.91	5.12	5.30	5.46	5.60
11	3.11	3.82	4.26	4.57	4.82	5.03	5.20	5.35	5.49
12	3.08	3.77	4.20	4.51	4.75	4.95	5.12	5.27	5.39
13	3.06	3.73	4.15	4.45	4.69	4.88	5.05	5.19	5.32
14	3.03	3.70	4.11	4.41	4.64	4.83	4.99	5.13	5.25
15	3.01	3.67	4.08	4.37	4.59	4.78	4.94	5.08	5.20
16	3.00	3.65	4.05	4.33	4.56	4.74	4.90	5.03	5.15
17	2.98	3.63	4.02	4.30	4.52	4.70	4.86	4.99	5.11
18	2.97	3.61	4.00	4.28	4.49	4.67	4.82	4.96	5.07
19	2.96	3.59	3.98	4.25	4.47	4.65	4.79	4.92	5.04
20	2.95	3.58	3.96	4.23	4.45	4.62	4.77	4.90	5.01
∞	2.77	3.31	3.63	3.86	4.03	4.17	4.29	4.39	4.47

From Reference 5.

the n's are all equal. If the n's are in fact not all equal but differ only slightly, replace n in the following discussion by the harmonic mean, n_H, of the n's,

$$n_H = t/(1/n_1 + 1/n_2 + ... + 1/n_t)$$

The procedure to be used is the following:

1. Choose a value for α, the significance level of the test (α is the risk of concluding that the averages differ when in fact they are the same).
2. Compute the variances of each mean, s_1^2, s_2^2, etc.
3. Compute $s_c^2 = 1/t(s_1^2 + s_2^2 + ... + s_t^2)$
4. Compute df for $s_c = (n_1 + n_2 + ... + n_t) - t$
5. Look up the value of $q_{1-\alpha}$ (t, df) in Table 9.3
6. Compute

$$w = \frac{q_{1-a}s_c}{\sqrt{n}}$$

7. If the absolute difference between any two means exceeds w (start with the largest minus the smallest), decide that the two respective averages differ, otherwise there is no reason to believe that they do.
8. Repeat until no differences exceeding w exist, in which case all remaining data are comparable.

TYPE I ERRORS, TYPE II ERRORS, AND STATISTICAL POWER

In the course of designing and planning measurement programs, it may be desirable to determine the minimum number of samples or measurements required to detect an important level of difference in them, with a given level of confidence. In such planning, it may be desirable to consider both the probability of false positive decisions (concluding that there is a difference when in fact there is none) denoted by α, and the probability of false negative decisions (failing to find a difference that actually exists) denoted by β. The required number of individuals (i.e., samples or measurements) will depend on the acceptable α and β risks, the magnitude of the difference that needs to be detected, and the standard deviation of the measurement process or sample population.

α = probability of saying there is a difference when in truth there isn't
β = probability of saying there isn't a difference when in truth there is
$1 - \beta$ = probability of saying there is a difference when in truth there is
$1 - \alpha$ = probability of saying there isn't a difference when in truth there isn't

When it is concluded that there is a difference when in in truth there isn't, it is said that a Type I error has occurred, the probability of which is usually denoted by α. When a difference that in truth exists has not been found, it is said that a Type II error has been committed, the probability of which is usually denoted by β. The quantity $1 - \beta$ is termed the statistical power of the test to detect a difference, d, with a sample of size n, when the test is carried at the α-level of significance when a difference actually exists. The situations described above are commonly summarized in the form

		Truth	
		Same	Different
Test	Same	Correct decision	Type II Error=β
Result	Different	Type I Error=α	Correct decision Power=$1-\beta$

The above chart shows the fundamental ideas behind hypothesis testing. In statistical theory, the statement 'Probability of saying there is a difference when in truth there isn't' is written as

$$Pr(\text{difference} \mid \text{no difference}).$$

This is read as 'Probability of saying there is a difference *given* there isn't'. If one denotes the null hypothesis, H_0, as there being no difference (same), and the alternative hypothesis, H_1, as there being a difference, then the above table may also be summarized as

		Truth	
		H_0=Same	H_1=Different
Test	H_0=Same	Correct decision $Pr(H_0\mid H_0)$	Type II Error=β $Pr(H_0\mid H_1)$
Result	H_1= Different	Type I Error=α $Pr(H_1\mid H_0)$	Correct decision Power=$1-\beta$ $Pr(H_1\mid H_1)$

Power, Type I, and Type II errors will have varying importance for the researcher depending on the situation. For example, an engineer may require a fluid to have a precise amount of a chemical in it due to safety concerns. He is interested in testing whether the amount of chemical present is safe. In this particular situation, it would be preferable to make a Type I error (conclude the fluid is unsafe when actually it isn't) to a Type II error (conclude the fluid is safe when actually it is not). It is better to discard a batch of good material than to potentially risk lives by using unsafe material. Further testing on the material declared unsafe may actually show its safety and thus the only waste would be duplication in testing. This is often the situation seen in mass screening programs, where a number of individuals are false positives but further testing reveals no problems. A great deal of thought into implications for Power, Type I, and Type II errors should be given when designing the experiment.

The expression that can be used to determine, n, the minimum number of samples or measurements that are required depends, in part, on whether or not the sign of the difference is important.

The Sign of the Difference is Not Important

If the sign is not important, which is to say that both a positive (+) and a negative (−) difference are of interest, the expression for use when σ is known or assumed (as contrasted with an estimated value), is

$$n = (z_{1-\alpha/2} + z_{1-\beta})^2 (s/d)^2$$

where d (here and throughout the following discussion) is the minimum detectable difference or the "important level of difference" mentioned at the beginning of this section.

When an estimate, s, of the standard deviation is used, the expression is

$$n = (t_{1-\alpha/2} + t_{1-\beta})^2 (s/d)^2$$

In the case of detecting a difference of a measured value, m, from a specified or standard value, m_0,

$$d = |m - m_0|$$

and σ or s = the standard deviation (or its estimate) of m.

In the case of detecting a difference between two measured values, m_A and m_B,

$$d = |m_A - m_B|$$

and

$$\sigma = \sqrt{(\sigma_A^2 + \sigma_B^2)}$$

or

$$s = \sqrt{(s_A^2 + s_B^2)}$$

Table 9.4, which is an abridgment of Table A.8 in Appendix A contained in Reference 5, may be used instead of the equations above to find the sample size (minimum number of samples or measurements) required to detect prescribed differences when the sign of the difference is not important. Reference 5 should be consulted if ratios of $d/\sigma < 1$ and/or other values for α are of interest.

The Sign of the Difference Is Important

There are times when the sign of the difference is important. That is to say, one is interested in whether a value exceeds a target value or whether a property of one product is larger than that of another (or, of course, the reverse situation) but not both.

Table 9.4. Sample Sizes Required to Detect Prescribed Differences between Averages when the Sign Is not Important

d/σ	α = .05 1 − β						
	.50	.60	.70	.80	.90	.95	.99
1.0	4	5	7	8	11	13	19
1.2	3	4	5	6	8	10	11
1.4	2	3	4	5	6	7	10
1.6	2	2	3	4	5	6	8
1.8	2	2	2	3	4	5	6
2.0	1	2	2	2	3	4	5
3.0	1	1	1	1	2	2	3

The expressions to be used to estimate n are similar to those in the previous section except

$$\text{substitute } z_{1-\alpha} \text{ for } z_{1-\alpha/2}$$
$$\text{or } t_{1-\alpha} \text{ for } t_{1-\alpha/2}$$

as the case may be. The values for the β terms remain the same.

Table 9.5, which is an abridgment of Table A-9 of Reference 5 may be used to find the sample sizes needed to detect differences of measured values when the sign of the difference is important. Again, Reference 5 may be consulted for situations not covered by Table 9.5.

Use of Relative Values

In the formulas in both the previous two sections, it is possible to use the relative standard deviation,

$$RSD = 100 \ (\sigma \text{ or } s/X)\%$$

and the relative difference

$$Rd = 100(d/X)\%$$

in the calculation of n, since the value of X cancels out in the arithmetic. Accordingly, one would substitute the ratio RSD/Rd in the expressions in the previous two sections. This procedure simplifies the calculations in many instances.

Table 9.5. Sample Sizes Required to Detect Prescribed Differences when the Sign of the Difference is Important

	$\alpha = .05$						
	$1 - \beta$						
d/σ	.50	.60	.70	.80	.90	.95	.99
1.0	3	4	5	7	9	11	16
1.2	2	3	4	5	6	8	11
1.4	2	2	3	4	5	6	9
1.6	2	2	2	3	4	5	7
1.8	1	2	2	2	3	4	5
2.0	1	1	2	2	3	3	4
3.0	1	1	1	1	1	2	2

The Ratio of Standard Deviation to Difference

The ratio σ/d, s/d, or the equivalent expression, RSD/Rd is very important in the design of measurement processes, especially since n (the minimum number of required samples or measurements) is directly proportional to the square of the ratio. If σ or s is small with respect to the difference that is to be detected, n can be relatively small. When σ or s approaches or exceeds the value of d, n becomes larger and it may be infeasible to look for a difference. In such a case, either a more precise method of measurement must be used (smaller σ or s) or one must be content with settling for a larger value for d.

Ideally, one would like to have the ratio

$$d/(\sigma \text{ or } s) = \text{or} > 10$$

Practically, a minimum value of 3 is quite reasonable. Such a value permits 95% confidence in the detection of d ($\alpha = \beta \approx 0.05$), with about one to two measurements, provided that the measurement system is in statistical control and that σ is known or s is based on at least 10 degrees of freedom.

Example It is desired to determine the minimum number of measurements that would be required to detect a relative departure of 10% of the effluent of a plant from its permitted discharge level with α and β probabilities of 0.05. The relative standard deviation of the measurement proposed is estimated to be 10%, based on 20 degrees of freedom.

Obviously, one is only interested in the probability of exceeding the permitted level (i.e., the sign is important). The calculation is as follows:

$$RSD/Rd = 10/10 = 1$$
$$t_{1-\alpha} = t_{1-\beta} = 1.725 \text{ for } \alpha = \beta = .05 \text{ and df} = 20$$
$$n = (1.725 + 1.725)^2 \times 1 = (2.975)^2 = 11.90 \approx 12$$

Questions concerning the ability to detect differences can be addressed, using the formulas given above or by using tables or graphs, based on the formulas. It has already been mentioned that Natrella includes in her book, *Experimental Statistics* [5], Table A.8 and Table A.9, which contain sample sizes required to detect prescribed differences between averages when the sign of the difference is, and is not, important, respectively. Sample sizes (minimum number of samples required) are given, depending on d/σ, α, and β.

Graphical solutions can be made using so-called OC (operating characteristic) curves. A typical OC curve, prepared for a given value for α, e.g., $\alpha = .05$, has values of d/σ for the abscissa and values of β for the ordinate. Such an OC curve is shown in Figure 9.1. Curves are traced for various values for n from which one can read the value for β corresponding for a selected number of measurements. Reference 5 contains a number of OC curves and describes their meaning.

CRITICAL VALUES AND P VALUES

The discussion surrounding statistical testing thus far has focused on critical values. This means that values of test statistics have been calculated and compared to critical values from various distributions with specific levels of probability and degrees of freedom. If the calculated value of the test statistic exceeds the critical value, the null hypothesis is rejected in favor of the alternative hypothesis. If the calculated value does not exceed the critical value, it is said that there is not enough evidence to reject the null hypothesis. Under this methodology, computer software such as MINITAB provides the test statistic value, and the user is left to find the appropriate table value for comparison.

The p value approach is slightly different. Although p value methodology is not focused on throughout this book, the reader should be aware of the principles surrounding p values. The p value approach tells the user the probability of seeing a more extreme value of the test statistic, given that the null hypothesis is true. Since a p value is a probability, it will be between zero and one. A very small p value (close to zero) indicates a small chance of seeing a more extreme result when the null hypothesis is true. This tells the user that there is a good chance that the altervative hypothesis is true. Conversely, a large p value indicates there is a good chance one could see a more extreme result even when the null hypothesis is true, telling the user there is a good chance that the alternative hypothesis is not true. Calculation of a p value depends on the whether the hypothesis being tested is two-sided or one-sided and if it is one-sided, the direction of interest.

MINITAB Example

Example   Consider the furnace temperature data from Chapter 5 for the situation of known standard deviation of 3. The hypothesis that was examined was

$$H_0 : \mu = 475 \text{ vs } H_A : \mu \neq 475 .$$

Intuitively, one should reject that the average population temperature differs from 475 degrees if a value of the test statistic (based on the sample mean) much larger or much smaller than 475 degrees was calculated. In terms of the standard normal probability distribution, this means we should reject for extreme values on both tail ends. With 95% confidence, this leaves 2.5% (0.025) for each tail region. Using standard normal tables, the rejection regions are values of the test statistic greater than 1.96 or less than –1.96. This is shown graphically in Figure 9.1. Now consider testing whether the average temperature is more than 475 degrees:

$$H_0 : \mu \leq 475 \text{ vs } H_A : \mu > 475 .$$

Intuitively, one should reject that the average population temperature is less than or equal to 475 degrees if a value of the test statistic (again based on the sample mean) much larger than 475 degrees was calculated. In terms of the standard normal probability distribution, this means we should reject for values on the right (upper) tail end. With 95% confidence, this leaves 5% (0.05) for the right tail region. Using

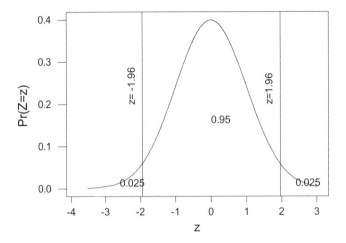

Figure 9.1. Critical regions for 2-sided hypothesis tests.

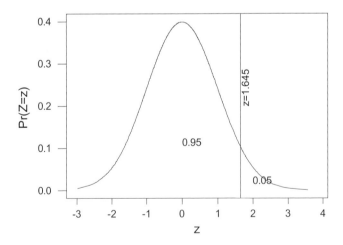

Figure 9.2. Critical region for 1-sided upper hypothesis tests.

the standard normal tables, the rejection region is values of the test statistic greater than 1.645. Notice how the positive rejection region is larger now than it was for the two-sided hypothesis because all probability is in the upper tail. This is shown graphically in Figure 9.2. Similarly, to test

$$H_0 : \mu \geq 475 \text{ vs } H_A : \mu < 475$$

one should reject that the average population temperature is greater than or equal to 475 degrees if a value of the test statistic (again based on the sample mean) much smaller than 475 degrees was calculated. In terms of the standard normal probability distribution, this means we should reject for values on the left (lower) tail end. With 95% confidence, this leaves 5% (0.05) for the left tail region. Using the standard normal tables, the rejection region is values of the test statistic less than -1.645. This is shown graphically in Figure 9.3.

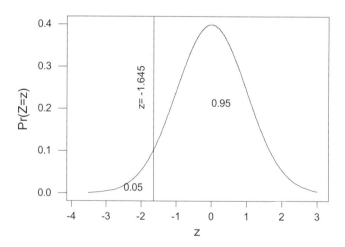

Figure 9.3. Critical region for 1-sided lower hypothesis tests.

These rejection regions are commonly used in the critical value approach to statistical testing. The p value approach calculates the probability in the tail end (for one-sided tests) or total probability in both tail ends (for two-sided tests) based on the calculated value of the test statistic (rather than the pre-defined critical value). The test statistic value for testing the average furnace temperature as 475 was z= -0.55 (shown in the MINITAB output in Chapter 5). This corresponds to the region

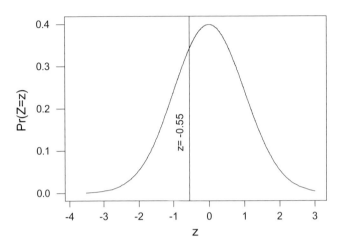

Figure 9.4. P value region.

shown in Figure 9.4. The p value is the probability under the normal curve from z= -0.55 to the left tail end. This area can easily be found using the following MINITAB menu commands:

Under Calc, Probability Distributions, Normal
- select **Cumulative Probability**
- enter 0 under **Mean**, 1 under **Standard Deviation**
- select **Input Constant** and enter –0.55, click OK

This provides the following output:

MINITAB Output

```
Cumulative Distribution Function

Normal with mean = 0 and standard deviation = 1.00000

        x      P( X <= x )
  -0.5500        0.2912
```

Taken from MINITAB

The interpretation is that the probability of a *smaller* value than the calculated test statistic of –0.55 is 0.2912. Thus for testing a one-sided lower hypothesis such as

$$H_0 : \mu \geq 475 \text{ vs } H_A : \mu < 475$$

one is interested in the probability the mean is less than 475, or alternatively, the probability the standard normal variable based on the Z is less than the standardized value of -0.55

$$\Pr(Z \leq -0.55) = 0.2912 .$$

Since this p value of 0.2912 is quite large (it implies there is a 29% chance one could see a smaller sample mean even if the true population mean was 475 or greater), one does not have evidence to reject that the true population mean is 475 or greater. If one was interested in the one-sided upper hypothesis

$$H_0 : \mu \leq 475 \text{ vs } H_A : \mu > 475$$

the probability of interest would be

$$\Pr(Z \geq -0.55) = 1 - 0.2912$$
$$= 0.7098$$

using the laws of probability defined at the beginning of the chapter. The p value tells the reader that there is now a 71% chance that a larger sample mean could be observed, even if the true population mean was 475 or less. Thus one does not have evidence to reject that the true population mean is 475 or less. Finally, to test the two-sided hypothesis

$$H_0 : \mu = 475 \text{ vs } H_A : \mu \neq 475 ,$$

the probability of interest is simply two times the p value for the one-sided lower hypothesis test

$$2 \Pr(Z \leq -0.55) = 2(0.2912)$$
$$= 0.5824.$$

It is helpful for the reader to calculate the p values for the furnace temperature data under the 3 hypothesis testing situations (one-sided lower, one-sided upper, two-sided) using MINITAB. The hypothesis test is controlled under the Options feature.

The general rules for calculating p values are as follows:

Hypothesis Test	p value
one - sided lower	$\Pr(z \leq z_c)$
one - sided upper	$\Pr(z \geq z_c)$
two - sided	$2\Pr(z \leq z_c)$

For general use in statistical testing, a p value of less than 0.05 is considered enough evidence to reject the null hypothesis. For p values 0.05 or greater, the null hypothesis is not rejected. P values provide information in terms of probability relating to how close one came to rejecting or not rejecting the null hypothesis. The p value of 0.2912 in the example is not at all close to 0.05, so that the user realizes the data was not even close to supporting the idea that the population mean was less than 475. All MINITAB output for statistical testing provides test statistic values as well as p values. The user can simply examine the p value, rejecting the null hypothesis only if the p value is less than 0.05. One can see what a powerful tool the computer is for statistical calculation.

CORRELATION COEFFICIENT

The correlation coefficient is used often as a measure of the association between two variables. Many statistical software packages include a program for such a calculation and the correlation coefficient, r, is routinely printed out in connection with other statistical parameters.

The correlation coefficient may be calculated by the expression

$$r = \frac{n \sum XY - \sum X \sum Y}{\sqrt{[n \sum X^2 - (\sum X)^2][n \sum Y^2 - (\sum Y)^2]}}$$
$$df = n - 2$$

where   X = value of the supposed independent variable
         Y = value of the supposed dependent variable
         n = the number of data pairs X,Y.

If   r =   0, no correlation of X and Y
            r =   +1, perfect positive correlation of X and Y
            r =   −1, perfect negative correlation of X and Y
            r =   intermediate values

An intermediate value can result because of variability considerations for the data. A value of zero is almost never obtained for r so that the question of whether a calculated value is significantly different from zero needs to be answered.

There are expressions for calculating confidence intervals for r depending on

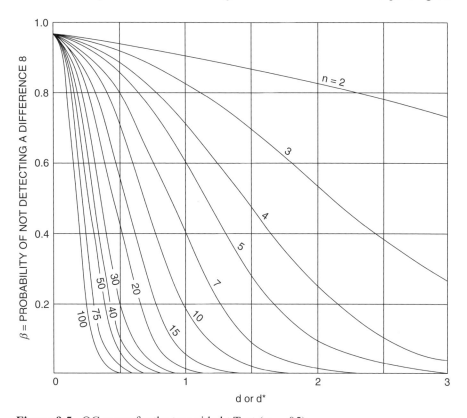

Figure 9.5. OC curve for the two-sided t Test ($\alpha = .05$).

sample size, n. Natrella includes in her book a graph (identified as Table A.17) from which such limits can be read [5]. The graph was prepared for decisions at the 95% confidence level. Table 9.6, based on Figure 9.5, illustrates the range of calculated r values that do not differ significantly from zero, based on sample size. Any value within the belt would not be considered to differ significantly from zero, with 95% confidence. Thus, for a data set of ten pairs, a value for r = 0.55 would not be considered to indicate positive correlation while a value of r = 0.65 would be so considered. Glass and Stanley present a general discussion of confidence intervals for correlation coefficients [6].

The square of the correlation coefficient, r^2, is called the coefficient of determination. It is defined as a measure of the amount of variation in the dependent variable that can be attributed to the independent variable. Thus, for $r = 0.65$, $r^2 = 0.42$ indicating that 42% of the variation in the dependent variable can be explained as due to the dependent variable. For $r = .98$, 96% of the variation could be so attributed. The coefficient of determination should be used cautiously as will be inferred from the ensuing comments about the correlation coefficient.

The correlation coefficient is often misunderstood [7]. A positive correlation simply means that Y is believed to increase when X increases. However, it must not be considered necessarily to indicate a causal relationship. There may be something

Table 9.6. 95% Confidence Belt for Correlation Coefficient

Sample size	df	Width of 95% Confidence Belt
3	1	- + .997
4	2	+ .95
5	3	+ .87
6	4	+ .81
8	6	+ .71
10	8	+ .63
20	18	+ .45
50	48	+ .33
100	98	+ .20
200	198	+ .14
500	498	+ .10

that causes both to change. A classical example is cited by Snedecor and Cochran [8] who point out that there is a very high negative correlation of −0.98 between the annual birthrate in Great Britain, from 1875 to 1920 and the annual production of pig iron in the United States! Also, there is a tendency to equate correlation coefficients with goodness of fit of data. The author has seen cases where a correlation coefficient of 0.997 was believed to be a better fit than 0.996 of a 5 point calibration curve. One can even find requirements in quality assurance plans to recalibrate if the correlation coefficient obtained is less than 0.995!

Natrella points out (cited in Reference 5) that it is meaningless to calculate a correlation coefficient if there is known to be a functional relationship, theoretical or empirical, between X and Y. This will often be apparent from a casual look at a graph of the data. It is when the data scatters markedly and when a graph is hard to draw that the correlation coefficient may begin to take on importance. Correlation coefficients should then indicate whether an investigation should be undertaken to look for causal relationships or other matters to understand the data. However, the experimenter should be very careful when drawing conclusions on the basis of these deceptively simple numbers [7].

Again, it is stated for emphasis that a residual plot is the best indicator of goodness of fit of experimental data. The statistical tests for randomness described earlier are the best means to decide whether a reasonable fit has been accomplished.

MINITAB Example

The correlation coefficient between X and Y may be calculated in MINITAB using the following menu commands:

Under Stat, Basic Statistics, Correlation
- double click on X and Y to add them to the **Variables** field, click OK

THE BEST TWO OUT OF THREE

An analyst or experimenter often is reluctant to make a decision on the basis of one observation or reading, and rightly so. (An exception to this is for the highly desirable case of knowledge of statistical control from other evidence than the data set in hand. Then, a single measurement can be relied upon with statistical confidence [9]) Two or more readings are frequently made not only to provide confidence that a blunder has not occurred but also to get some idea of the precision of measurement when this is the only information available. Some analysts have been known to practice, even up to the present time, the procedure of taking three measurements and discarding the most remote value, that is to say the best two out of three are retained. Youden [10] has shown that this practice increases the apparent variability of replication rather than improving it.

To demonstrate the last statement, Youden made drawings of triads from a large group of measurements, known to be normally distributed and free from gross errors. Then, he calculated the ratio of the larger difference to the smaller difference between the data in the triads. Thus, in the triad 10, 14, 15, the ratio would be 4 to 1. He found large ratios to occur more frequently than expected by the nonstatistician. In 400 sets, a ratio of 4.0 or more occurred 149 times, a ratio of 9.0 or more occurred 70 times, and 19.0 or more occurred 38 times. Thus, on the average, one out of twelve sets was 19 times farther away than the distance separating the two closest ones. Accordingly, using the best two out of three rule results in discarding good data and exaggerates the idea of the precision that is obtained.

It can be shown further that when two readings are taken, say 12 and 15, the third reading is likely to be closer to either the smaller or the larger, rather than in the middle of the range. One can see that data rejection would only increase the variability of the means of a number of such sets.

COMPARING A FREQUENCY DISTRIBUTION WITH A NORMAL DISTRIBUTION

The frequency distribution of a data set may be compared with a normal distribution by superimposing a normal curve upon it. A procedure that can be used is the following:

1. Prepare a frequency distribution for the data set as described in Chapter 7.
2. Compute $\overline{X}$ and s of the data set.
3. Estimate a value for $\overline{Y}$, the frequency corresponding to $\overline{X}$. (Some judgment may be required in determining $\overline{Y}$. It may be the experimental frequency obtained from the plot, or some lesser or greater value that appears to be a better fit of the plot).
4. Calculate values for Y corresponding to selected values of X, using the format given in Table 9.7. Note that the values for 'NF', the normalization factor, are obtained from Table 9.8.
5. Sketch the normal distribution curve on the same graph paper that the frequency distribution was drawn.

The values suggested for k in Table 9.7 are adequate to sketch the normal curve for

Table 9.7. Format for Use in Construction of a Normal Distribution

Multiple of Standard Deviation k	X Values		$Y = \overline{Y}$ ("NF")
	$\overline{X} - k\,s$	$\overline{X} + k\,s$	
0			
0.5			
1.0			
1.5			
2.0			
2.5			
3.0			

most purposes. Intermediate values for k may be used if desired to draw a more accurate normal curve.

A series of 31 measurements ranged in value from 10.02 to 10.34. After coding the data by subtracting 10.00 from each value, the resulting differences were multiplied by 100 for ease of computation. The coded data were put into classes, as shown, for construction of the frequency histogram.

Class	0-5	5-10	10-15	15-20	20-25	25-30	30-35	35-40
Number	2	0	2	6	10	8	3	0

$$\overline{X} = 22.5 \quad s = 7.0 \quad \overline{Y} = 10$$

The frequency histogram is shown in Figure 9.6.

The normal curve for superposition on the frequency histogram is constructed from the following data.

k	$\overline{X} - ks$	$\overline{X} + ks$	Y = 10 "NF"
.5	19.0	26.0	8.8
1.0	15.5	29.5	6.1
1.5	12.0	33.0	3.2
2.0	8.5	36.5	1.4
3.0	1.5	43.5	.1

Table 9.8. Normalization Factors for Drawing a Normal Distribution

Multiple of Standard Deviation, k	"NF"
0.00	1.000
.25	.965
.50	.883
.75	.755
1.00	.607
1.25	.458
1.50	.325
1.75	.216
2.00	.135
2.25	.079
2.50	.044
2.75	.023
3.00	.011

The superimposed normal curve appears to be a fair representation of the data.

CONFIDENCE FOR A FITTED LINE

A confidence band for a fitted line as a whole describes the area in which lines fitted to repeated experiments of the same kind are expected to lie. Each set of experiments would produce a somewhat different set of pairs of X, Y values and, accordingly, a different linear fit would result. The confidence band defines the expected spread of a large set of such lines. Thus, a 95% confidence band would define the region within which 95 out of 100 repetitive lines would be expected to lie.

The confidence band is defined by an upper and a lower hyperbola the foci of which are located at the midrange of the fitted line. The computations necessary to obtain a least-squares fit (required) provide the information to construct the confidence band. The equations for the upper and lower branches of the hyperbola are as follows [11]:

Upper branch

$$Y_u = a + bX + (2F)^{1/2} s_Y [\frac{1}{n} + \frac{(X - \overline{X})^2}{S_{xx}}]^{1/2}$$

Lower branch

$$Y_l = a + bX - (2F)^{1/2} s_Y [\frac{1}{n} + \frac{(X - \overline{X})^2}{S_{xx}}]^{1/2}$$

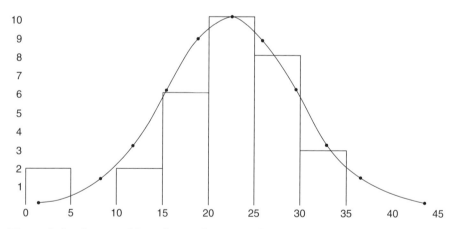

Figure 9.6. Superposition of normal curve on frequency plot.

The values for a and b come from the least squares fit. The values for s_Y and S_{xx} are obtained from the least squares worksheet (see Table 7.5 in Chapter 7) or can be calculated as shown on the worksheet. The value for F depends on the confidence level sought for the band, i.e., $(1 - \alpha)$, and the number of degrees of freedom for the fit of the line. More precisely, look up $F_{1-\alpha}$ for $(2, n - 2)$ degrees of freedom.

Ordinarily, selected points are calculated and the branches of the hyperbola are sketched. The midpoint, two points near each end of the line, and two points equispaced between them are computed for this purpose. The computation procedure is as follows:

1. The values for X and Y at the center are $\overline{X}$ and $\overline{Y}$ from the least squares computation. Values for Y_c at selected values of X are calculated from the equation of the fitted line, $Y_c = a + bX$.
2. Obtain s_Y and S_{xx} from the work sheet.
3. Look up F_{1-af} for (2, n − 2) for $1 - a = .95$ form the selection found in Table 9.9. (For values not listed here or for other than 95% confidence level, consult other compilations of statistical tables, e.g., Reference 5.)

Mandel and Linnig [12] describe a graphical method of constructing the confidence band hyperbola in a general discussion of the reliability of a calibration line.

From an inspection of the equations for the branches of the hyperbola, it is evident that the F value for a line fitted to three points is meaningless, and that at least five points are needed to have any statistical confidence in a fitted line. (This is because of the large value of F that would be required, reflecting the large uncertainty of estimates of standard deviations, based on a few degrees of freedom.) It will be noted, as well, that the width of the confidence band is directly proportional to the standard deviation of the fitted points, s_Y. For every imprecise methods of measurement (large scatter of points around the line), large variability between repetitive fits would be expected. This can be compensated somewhat by an increased number of data pairs, n, but this is not a direct proportionality, as seen from the equation.

Table 9.9. Values for $F_{1-\alpha}$ (α = .95) for (2, n − 2)

n-2	1	2	3	4	5	6	7	8	9	10
F(2,n-2)	199	19	9.55	6.94	5.79	5.14	4.74	4.45	4.26	4.10

11	12	13	14	15	20	25	30	40	60	120	∞
3.98	3.89	3.81	3.74	3.68	3.49	3.39	3.32	3.23	3.15	3.07	3.00

Example   The calibration experiment will be presented to illustrate the construction of confidence bands.

Calibration data were fitted by the method of least squares to a straight line in Chapter 7. The equation obtained was

$$Y_c = 0.161 + 10.0833 \, X$$

and

$$S_{xx} = 60 \quad s_Y = 2.2304$$

Also

$$F_{.95} = 4.74$$

hence

$$(2F)^{1/2} = 3.079$$

When these quantities are substituted in equations for the confidence hyperbolas, one obtains

$$Y_u = 0.161 + 10.0833 + 3.079 \times 2.2304 \ (1/9 + (X - \overline{X})^2/60]^{1/2}$$

and

$$Y_1 = 0.161 + 10.0833 - 3.079 \times 2.2304 \ [1/9 \ (X - \overline{X})^2/60]^{1/2}$$

These may be rewritten

$$Y_u = Y_c + Q \text{ and } Y_1 = Y_c - Q$$

Calculating Y_u and Y_1 for several values of X, one obtains

X	Y_c	Q	Y_u	Y_1
1	10.24	4.22	6.02	14.46
3	30.41	2.90	27.51	33.31
5	50.58	2.29	48.29	52.87
7	70.74	2.90	67.84	73.64
9	90.91	4.22	96.69	95.13

The reader may want to plot the data and the fitted line and construct the 95% confidence bands. Intermediate points may be calculated if there is difficulty in constructing the hyperbolas.

MINITAB Example

MINITAB has the capabilities to produce the confidence regions around a fitted line. After entering the calibration data into MINITAB, the following menu commands produce the fitted line plot in Figure 9.7 with 95% confidence bands:

Under Stat, Regression, Fitted Line Plot
- double click on Y to add it to the **Response** field and X to add it to the **Predictor** field
- select a linear regression model
- under **Options**, select Display Confidence Interval, enter 95 in the **Confidence Level** field and a title in the **Title** field, click OK, click OK

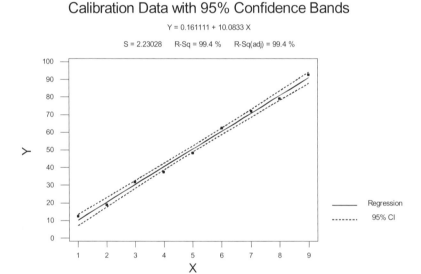

Figure 9.7. Calibration data with confidence bands.

JOINT CONFIDENCE REGION FOR THE CONSTANTS OF A FITTED LINE

In the section Confidence for a Fitted Line, the confidence band for a fitted line was discussed. In the present section, the confidence region for the constants for a fitted line will be considered.

The 95% confidence interval for the intercept, a, and for the slope, b, of a fitted line, on first consideration, might be expected to be a ± 2 s_a and b ± 2 s_b, respectively. However, this is not the case, because the intercept and slope are correlated for any given line. Accordingly, all of the values within each interval are not mutually compatible. When all consistent intercepts and slopes are plotted on a graph, with the slope as abcissa and the intercept as ordinate, the area is not bounded by a rectangle but rather by an ellipse, as shown in Figure 9.8, which consists of a computer-drawn plot, using the data of Chapter 7.

The equation for the joint confidence ellipse given by Mandel and Linnig is presented in [12]:

$$2Fs^2 = N(a-a')^2 + 2\sum X(a-a')(b-b') + \sum X^2(b-b')^2$$

where   F = critical value for the variance ratio with 2 and N-2 degrees of freedom
            a' = the value for the intercept of the fitted line
            b' = the value for the slope of the fitted line
            s = the estimate of the standard deviation of the Y measurements
            N = the number of pairs of X, Y values used to fit the line

Mandel and Linnig [12] also describe a graphical way to construct the joint confidence ellipse.

The joint confidence ellipse is the loci of points that would be expected to contain the true values for the constants a and b of the fitted time, with the specified confidence. It may also be used to decide whether the constants obtained for the fit of a similar set of data at some other time were consistent with an earlier set. An extension of this is its use in judging the consistency of calibration lines obtained on various occasions, and even of calibration lines obtained by several laboratories.

SHORTCUT PROCEDURES

There are a number of shortcut tests, characterized by their simplicity, that statisticians have devised for hypothesis testing. They are easy to apply and require only the requisite tables and a simple calculator. Their main disadvantage is that they have less statistical power, i.e., a larger value for β, than the tests described in Chapter 5. All of these tests require that the data are normally distributed. These tests are described in detail in Chapter 15 of Reference 5, together with examples.

In answering the question, "Does a measured value differ from a standard value?" (see Chapter 5) the difference found is divided by the range of the measured values (i.e., the largest minus the smallest value) and this quotient is compared with a table value that depends on the significance level of the test and the number of measurements. If the quotient exceeds the table value, conclude that the difference is significant at the level of the test.

In answering the question, "Do two means differ significantly?", the difference of the two measured values is divided by the average of the ranges of the two sets of measured values from which the respective means were calculated. This quotient is compared with a tabulated value that depends on the significance level of the test and the number of measurements on which each mean is based (each based on the same number). If this quotient exceeds the table value, it is concluded that the two means differ. The procedure is slightly different if pairs of measurements are involved (see Reference 5). The tests described above may be conducted according to whether the sign of the difference is or is not important.

Two products (or measurement methods) may be compared for variability by a shortcut method, similar to the F test. In this test, the ratio of the respective ranges of two sets of measurements is computed and designated F'. This ratio is compared with a tabulated critical value for F' that depends on the number of measurements for each set of data, and the significance level chosen for the test.

NONPARAMETRIC TESTS

These tests relate to situations where the populations of interest are not normally distributed. One such test is called the Wilcoxon Signed-Rank test.

Wilcoxon Signed-Rank Test

The Wilcoxon Signed-Rank test involves ranking the data and assigning signs to the rank numbers. The sum of the positive ranks is compared with a table value C, that depends on the significance level of the test and the number of observations in the data set. This may be written as

$$\boxed{S_+ = \text{sum of ranks associated with positive}\,(x_i - \mu_0)}$$

where

$$\mu_0 = \text{the hypothesized value of the mean under the null hypothesis}$$

The null hypothesis

$$H_0 : \mu = \mu_0$$

is rejected according to the following:

Alternative Hypothesis	Rejection Region
$H_1 : \mu \neq \mu_0$	$S_+ \geq C, S_+ \leq \dfrac{n(n+1)}{2} - C$ at $\alpha/2$
$H_1 : \mu > \mu_0$	$S_+ \geq C$ at α
$H_1 : \mu < \mu_0$	$S_+ \leq \dfrac{n(n+1)}{2} - C$ at α

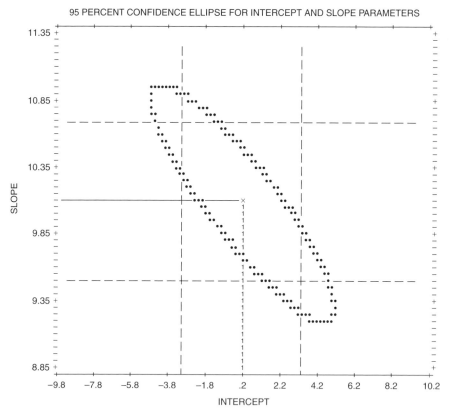

Figure 9.8. Joint confidence region ellipse for slope and intercept of a linear relationship.

In using this test, the data are assumed to be symetrically distributed about a mean value. Table values are provided in [13].

Example Consider the calibration data from Chapter 7. The metrologist would like to test with 99% certainty whether the calibration mean is 64.0 and is only interested in deviations below. The hypothesized mean, 64.0, is subtracted from each original data point Y. The absolute values are taken, and the data ranked in ascending order to give Table 9.10:

Table 9.10. Wilcoxon Signed-Rank Test Calculations

Order of Absolute Magnitude	Rank	Sign
1.5	1	-
8	2	+
15	3	+
15.8	4	-
26.5	5	-
28.5	6	+
32	7	-
45	8	-
51.5	9	-

Then

$$S_+ = 2+3+6$$
$$= 11.$$

Using the appropriate entry in Table A.9 of [13], for 99% confidence (i.e., alpha=0.01),

$$\Pr(S_+ \geq 42 \text{ when } \mu_0 = 64) = 0.01$$

for a sample size of 9. Since

$$H_1 : \mu < 64$$

the critical rejection region is

$$S_+ \leq \frac{9(10)}{2} - 42$$
$$= 3$$

so that

$$S_+ = 11 > 3$$

and the metrologist concludes there is no reason to reject a mean calibration of 64 in favor of one lower.

The Wilcoxon Signed-Rank test may be applied to comparing the averages of several products. More details are given in Reference 5 which also contains examples.

MINITAB Example

Although not computationally difficult, the Wilcoxon Signed-Rank test may be tedious for large datasets. MINITAB offers several non-parametric procedures including the Wilcoxon Signed-Rank test. The following menu commands provide the analysis for the above data:

Under Stat, Nonparametrics, 1 Sample Wilcoxon
- double click on Y to add it to the **Variables** field
- enter 99 in the **Confidence Interval** field
- enter 64 in the **Test Median** field and select **Alternative** less than, click OK

The MINITAB output is shown below.

EXTREME VALUE DATA

The average value of a set of data is the parameter used in many applications.

MINITAB Output

```
Wilcoxon Signed Rank Test: Y

Test of median = 64.00 versus median  <  64.00

                N for   Wilcoxon              Estimated
          N     Test    Statistic       P      Median
Y         9       9         11.0     0.096      50.00
```
Taken from MINITAB

However, there are cases in which the smallest or the largest value is the one of most concern. For example, metrologists may be interested in the lowest temperature in a given period of time or lengths of periods of drought.

Material scientists may be interested in maximum or in minimum strengths of materials. The extreme values, e.g., the maximum temperature in a month, may be ranked in order of size, from the smallest to the largest, as

$$X_1 < X_2 \ldots < X_i \ldots < X_n$$

A plotting position, P, is calculated according to

$$P = i/(n-1)$$

where i is the rank of the point to be plotted. A plotting paper is used in which the scale of the X-axis (P) is nonlinear (actually a double exponential scale) while the Y-axis (property) is linear.

Figure 9.9 is included to show the kind of plotting paper that is used and to illustrate the technique. The data plotted consist of maximum values obtained on the tensile strength (in psi) for an aluminum alloy over a number of heats. A reasonable straight line plot of the data indicates that extreme value statistics may be applied. Accordingly, one could infer that a tensile strength of 33,500 psi would be found once in 100 heats.

Extreme value techniques are discussed more fully in Chapter 19 of Reference 5.

STATISTICS OF CONTROL CHART

Before concluding this chapter, it may be useful to summarize the statistical techniques that are especially useful in developing and utilizing control charts. The techniques mentioned are limited to Shewhart control charts which are the ones most frequently used in measurement laboratories. It is assumed that the reader has general knowledge of such charts. For further information, the reader may wish to consult the author's book on quality assurance [9].

Property Control Charts

Property control charts denote those in which a single measurement, X, or the mean of several measurements, X_1, of a "control sample" such as a stable test sample, a blank, a spike, or even an operational parameter of a system is plotted as a function of sequence or time. A "central line" indicates the expected value of the measure-ment, while upper and lower warning limits indicate the range within which the measured values are expected to lie, 95% of the time. "Control limits" are drawn to indicate the range within which virtually all of the values should lie when the measurement system producing them is in statistical control. The limits are calculated from estimates of the long-term standard deviation, s_b, based ideally on df > 15, but never on df < 7.

Table 9.11. lists the basis for calculation of the control chart limits. The table also includes values for calculating the limits of range charts which will be discussed later.

The central line can be either the known value for the control sample or a value resulting from its repetitive measurement.

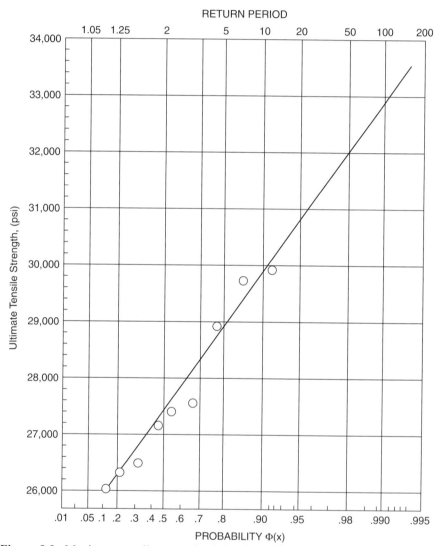

Figure 9.9. Maximum tensile strength of aluminum alloy.

Control chart limits need to be reevaluated from time to time and revised as necessary. If the central line is based on repetitive measurements of the control sample, one can decide whether the mean of a later set of measurements is significantly different from an earlier set by using the technique described in Chapter 5. Testing at the 95% level of confidence is recommended. Should the central line value prove to be significantly different, the later determination is the one that is preferred to be used for the central line. If not, the means should be combined to give one with greater confidence, using the techniques described in Chapter 6.

Table 9.11. Control Chart Limits

Type of Chart	Central line	WL	CL
X Chart	$\bar{X}$ or known value	$\pm 2\, s_b$	$\pm 3\, S_b$
$\bar{X}$ Chart	$\bar{X}$ or known value	$\pm 2\, s_b/\sqrt{n}$	$\pm 3\, s_b/\sqrt{n}$
R Chart*			
Duplicates	$\bar{R}$	+ 2.512 R	+ 3.267 R
Triplicates	$\bar{R}$	+ 2.050 R	+ 2.575 R
Quadrupli cates	$\bar{R}$	+ 1.855 R	+ 2.282 R
Pentuplicates	$\bar{R}$	+ 1.743 R	+ 2.115 R
Sextuplicates	$\bar{R}$	+ 1.669 R	+ 2.004 R

* Control limits shown are upper limits. LWL and LCL = 0 in each case.

When a stable test sample of known properties is used as the control sample, the known value is the one used for the central line, provided that the test data do not differ significantly from it. To make this decision, the mean of a sufficient number of measured values is compared with the known value, using the technique described in Chapter 5. If there is a significant difference, the measurement system is producing biased results and its malfunction needs to be investigated.

To test for the significance of apparent changes in s_b, use the F test described in Chapter 5. Again, if no significant change is believed to have occurred, the estimates should be combined using the pooling technique described in Chapter 4.

Otherwise, the latest estimate of the standard deviation is the one that should be used.

Precision Control Charts

The range of a series of measurements may be plotted on a control chart to monitor the precision of a measurement system. The absolute value of the range, i.e., the largest minus the smallest value is used for the plot. The "central line" is the average. $\bar{R}$, of the ranges found when measuring at least 7 and preferably at least 15 sets of test samples. Often the test samples may be the actual laboratory samples that are measured.

It should be remembered that a range chart, called an R chart, only monitors precision. In the case of actual laboratory samples, groups of samples may be represented on the control chart, providing their properties do not differ too much as was discussed in Chapter 4.

The factors by which $\bar{R}$ is multiplied to compute the control limits are given in Table 9.7. The measurement system is considered to be in control, precision wise, as the plotted values of ranges fall within the control limits. The simplest way to decide whether $\overline{R}$ has changed during the course of use of a control chart is to convert the values to be compared to standard deviations and then to apply the F test. The procedure to be used in revising limits is the same as discussed above.

A modification of the range chart to extend its usefulness to a large span of property values is called the "relative range chart". The technique for its preparation and rational for its use are described in Reference 9.

Systematic Trends in Control Charts

In addition to remaining within control limits, plotted points should not exhibit systematic trends. The user should look for apparent departures from randomness, using the procedures described earlier in this chapter. The techniques that should be most useful include the test for runs, the test for trends and slopes, and the mean squares of successive differences test. Some computer programs are capable of making some of these tests on a routine basis.

By keeping control charts in essentially real time, statistically analyzing the results, and keeping good collateral records, it may be possible to diagnose the causes of measurement system malfunctions and to take appropriate corrective actions to stabilize and even improve them.

SIMULATION AND MACROS

Simulation is a powerful technique which enables users of statistics to examine the outcome when an experiment is performed many times (hundreds or even thousands) using the computer. A macro is the computer code that tells the computer how to conduct the simulation. Thus macros cannot be generated using the menu commands that have been the cornerstone of the analyses presented so far. The user must actually understand the computer language for the software to write a macro. This is often the most challenging part of using macros! One must be able to translate what the conditions of the experiment are into the software's language. The specific terms of the experiment (including type of data and number of times the experiment is to be conducted) are included in the macro. Macros use random numbers to generate many samples of data (all based on the same population) to examine the properties of the samples. Simulation is often used for generating sample sizes and looking at power in situations where sample size/power formulas are not available. For the most part, simulations are only limited by the knowledge of the user in terms of conditions of the experiment and expertise in writing macros. The general procedure in MINITAB is that the macro itself is a text file that is stored elsewhere. A command is then given in MINITAB to invoke or call the macro.

MINITAB Example

Example   Consider the example presented in Chapter 6 where a metrologist wanted to randomly generate 100 observations representing the amount of air toxin, knowing that the data was normally distributed with mean 10.295 micrograms and standard deviation 2.269 micrograms. The Random Data menu in MINITAB easily generated this data. However, what if the metrologist wanted to generate these 100 observations 500 times to examine the 500 means from each sample? She could certainly repeat the menu command 500 times, each time using menu commands to calculate the mean. This is not an efficient use of time, so she decides to use a MINITAB macro to perform the calculations. It is helpful to first examine the MINITAB commands used to generate the 100 observations, without regard to the simulation. MINITAB commands can be viewed in the Session Window by selecting the menu commands Editor, Enable Commands. The cursor must be in the Session Window to select these menu commands. Now for every menu command selected, the actual MINITAB code will appear in the Session Window. The menu commands used previously to generate the 100 observations

Under Calc, Random Data, Normal
- enter 100 in the **Generate Rows of Data** field
- enter C2 in the **Store in Columns** field
- enter 10.295 in the **Mean** field and 2.269 in the **Standard Deviation** field, click OK

produce the MINITAB code in the Session Window:

MINITAB Session Window

```
MTB > Random 100 c2;
SUBC>    Normal 10.295 2.269.
```
Taken from MINITAB

The first line tells MINITAB to generate 100 random numbers and put them in column C2. The subcommand line says use the Normal distribution with mean 10.295 and standard deviation 2.269. The next step is to use a macro to repeat this process 500 times, each time calculating the mean. The macro itself is saved as a text file in word processing software. The macro is called at the MINITAB prompt when the user wants it to be executed. Below is the macro code that performs the desired operation, with comments (in italic) regarding the purpose of each line. Note that the italicized comments are not actually part of the file and are only shown here for comprehension.

```
Macro                   -starts the macro
Means x y               -names the macro means and defines x  and y as
                         arguments
mcolumn x y             -defines macro columns x and y
mconstant I             -defines macro constant I
```

```
do I=1:500              -tells the macro to start a loop
                        which will be performed 500 times
random 100 x;           -generate 100 random observations and put them
                        in macro column x
normal 10.295 2.269.    -follow the normal distribution
                        with mean 10.295, standard deviation 2.269
let y(I)=mean(x)        -take the mean of column x and store it in row
                        I of column y
enddo                   -end the loop
endmacro                -end the macro
```

The text above should be saved in a text file (extension txt) and stored in the MINITAB macro directory (this is c:\program files\mtbwin\macros as the default). This file was saved as 'means.txt'. The macro is then invoked using the following menu commands at the command prompt:

MINITAB Session Window

```
MTB > %means.txt c1 c2
```

Taken from MINITAB

This tells MINITAB to start the macro entitled 'means', using C1 to store the random numbers and C2 to store the means. Note that C1 is replaced with 100 new random numbers every time the loop is performed. Essentially, this macro performs a simulation to find the 500 means resulting from 500 normally distributed samples.

The author speaks from experience in saying that the major hurdle in macro programming is becoming familiar with the language. A good starting point is to try and find a macro which performs a similar operation to that desired and modify it step by step. Many macros can be found in the macro directory of MINITAB. Do not make many changes to a macro all at once. It is preferable to make one change at a time, making sure the macro still runs after each change. Macros are a very powerful and rapid investigative technique. More information on macros can be found under Help, Session Command Help, Using Macro Commands from the MINITAB menu commands or in [14].

EXERCISES

9-1   What is the probability of getting a 6 on casting a six faced die? What is the probability of getting any number but 6?

9-2   What is the probability of drawing an ace of clubs from a well shuffled deck of cards?

9-3   What is the probability of drawing four aces successively from a deck of well shuffled cards?

9-4   Calculate and compare the 25% trimmed mean of the data set given with the arithmetic mean, and the median of the data set:

11.1, 12.0, 11.5, 11.3, 11.1, 11.9, 12.3, 13.0, 12.7, 11.6, 12.3, 13.1, 12.8,
13.5, 11.9, 11.7, 12.4, 13.4, 12.9, 13.0

9-5   Plot the data of exercise 4 and comment on any anomalies.

9-6   Plot the data of exercise 4. Find the least squares linear slope and decide on its significance.

9-7   Calculate the expected number of runs and compare with the number experimentally found. What significance do you attach to these calculations?

9-8   Calculate the MSSD for the data set of exercise 4 and use it to evaluate the randomness of the data.

9-9   Five measurements were made on the moisture content (in %) of each of five products with the following results:

A, 5.34, s = 0.202; B, 5.36, s = 0.240; C, 5.63, s = 0.252;
D, 5.49, s = 0.126; E, 5.27, s = 0.307

Do the individual products differ significantly at the 95% level of confidence? Do the variances differ significantly? Calculate a grand average and the 95% confidence interval for it.

9-10 In planning a measurement program, it was decided that a relative departure of ± 0.50% had to be detectable with 95% confidence. Based on duplicate measurements, what would be the maximum relative standard deviation the could be tolerated in the measurement process? If the best attained precision were 0.25%, how many measurements would be required to monitor the process for compliance with the requirement?

9-11 The following data consist of single measured values of a spiked sample (spiked at 0.12 mg/L) reported by 24 laboratories in a laboratory proficiency test

Lab No.	Result	Lab. No.	Result
1	.10	13	.13
2	.121	14	.115
3	.07	15	.097
4	.047	16	.18
5	.12	17	.14
6	.106	18	.382
7	.100	19	.14
8	.32	20	.11
9	.13	21	.09
10	.099	22	.11
11	.097	23	.12
12	.137	24	.11

a. Rank the data in ascending order.
b. Plot the ranked data, and note whether a symmetrical plot is achieved, indicating essential normality.
c. Determine the median value.
d. Calculate the 25% trimmed mean.
e. Estimate the standard deviation, based on the 25% Winsorized range.
f. Decide whether the mean is or is not significantly biased.
g. Decide which labs reported results outside of the 95% confidence bands.
h. Decide whether any labs reported results outside of the 99.7% confidence bands.
i. Using the Wilcoxon Signed-Rank test, determine whether the mean result is 0.20 with a two-sided rejection region.

9-12. For the MINITAB data set 'plating.mtw', use the Wilcoxon test to determine whether the average number of parts with a plating defect is 4. Assume that only departures greater than 4 are of interest.

9-13. Write a MINITAB macro to generate 1000 samples of 200 each from a Binomial distribution with probability of success p=0.25.

REFERENCES

[1]   Eisenhart, C., "The Meaning of 'Least' in Least Squares," *J. Wash. Acad. Sci.*, 54: 24-33 (1964).

[2]   Crow, E.L., and M.M. Siddiqui, "Robust Estimation of Location," *J. Am. Stat. Assoc.*, 62: 353-89 (1967).

[3]   Ku, H.H. "A Users' Guide to the OMNITAB Command "STATISTICAL ANALYSIS," NBS Technical Note 756, (Gaithersburg, MD: National Institute of Standards and Technology, 1973).

[4]   Hart, B.I., "Significance Levels for the Ratio of the Mean Squares of Successive Differences to the Variance," *Ann. Math. Statist.*, 13: 445-447 (1942); see also, Sachs, L *Applied Statistics, A Handbook of Techniques* (translated by Z. Reynarowych), (New York: Springer-Verlag 1982).

[5]   Natrella, M.G., "Experimental Statistics," NBS Handbook 91, (Gaithersburg, MD: National Institute of Standards and Technology 1963, p 3-40).

[6]   Glass, G.V., and J.C. Stanley, *Statistical Methods in Education and Psychology* (Gaithersburg, MD: Prentice-Hall, Inc., 1970).

[7]   Arni, H.T. "The Significance of the Correlation Coefficient for Analyzing Engineering Data," *Materials Res. Stand.*, 11: 16-19 (1971).

[8]   Snedecor, G.W., and W.G. Cochran, *Statistical Methods* 6th Edition, (Ames, IA: Iowa University Press 1967, Sect. 7.8).

[9]   Taylor, J.K., *Quality Assurance of Chemical Measurements* (Chelsea, MI: Lewis Publishers, Inc., 1987).

[10]   Youden, W.J., "The Fallacy of the Best Two Out of Three," *Tech. News Bull. Natl. Bureau Standards*, July, 1949.

[11] Youden, W.J., "Experimentation and Measurement," NBS Special Publication 672, (Gaithersburg, MD: National Institute of Standards and Technology, 1984).

[12] Mandel, J., and F.J. Linnig, "Study of Accuracy in Chemical Analysis Using Linear Calibration Curves," *Anal. Chem.*, 29: 743-749, (1957).

[13] Devore, J.L. *Probability and Statistics for Engineering and the Sciences* (Monterey, CA: Brooks/Cole Publishing, 1987).

[14] *MINITAB User's Guide 1: Data, Graphics and Macros*, Release 14 for Windows (Minitab Inc. 2003).

CHAPTER 10

Conclusion

SUMMARY

The statistical techniques that have been presented in this book are applicable to many of the situations that are encountered when dealing with ordinary measurement data. They are equally useful whether one is dealing with chemical measurements, or physical, or engineering data. They are, of course, based on the assumption that one has data that are worthy of statistical evaluation. That is to say, that data should be the output of a system under statistical control and should be reproducible and representative of what would be expected if that same set of measurements were undertaken again under the same conditions of measurement. The statistical techniques that have been presented provide the tools for deciding whether that is the case when replicate measurements are made.

As with any discipline, thought and understanding play a large role in becoming knowledgeable in that field. It has been stated previously in this text, but it is worth stating one final time that good statistics applied to poor data always give poor results. Poor data can result from poor design, execution or measurement of an investigation. By the same token, poor statistics applied to good data also give poor results. Obviously an appropriate statistical test must be chosen. Another common error is to keep analyzing the data until the desired result is realized. This is referred to as 'data dredging' in the statistical literature. The importance of specifying what one wants to look at in advance is paramount. It is permissible to change one's mind on this during the experiment, but not at the final stage during data analysis! Although not everyone wants to acknowledge it, a negative result is still important because it provides clarification. Finally, keep in mind that although a result may be statistically significant, it may not have real-world meaning. Conversely, a result may not reach statistical significance but have important implications.

Now the reader is urged to use the tools of statistics. The more one uses them, the clearer the concepts will become and the greater will be the facility gained in the ability to apply them. Therefore, the author's plea to the reader is to understand the principles that are involved. Rework all of the examples in the book to verify understanding. Complete all of the exercises and file the results in an appropriate place for future reference. Use the formulas provided but do not just substitute values in them. As for statistical software, don't forget that the output is only as good as the knowledge of the user. Make an effort to understand what is being calculated and attempt the calculation yourself (even with a small dataset) before producing the calculation with software. Arriving at a well-understood, correct answer is the most rewarding part of the learning process.

Statistical Tables

This Appendix contains a selection of statistical tables for use in solving the problems in the text. The tables are abbreviated forms of larger tables found in statistical books, such as National Bureau of Standards Handbook 91, referred to in several places throughout the text.

Table A.1. Use of Range to Estimate Standard Deviation

Number of Sets of Replicates	Factor Degrees of Freedom	Number of replicates in set				
		2	3	4	5	6
1	$d_2^;$	1.41	1.91	2.24	2.48	2.67
	df	1.00	1.98	2.93	3.83	4.68
3	$d_2^;$	1.23	1.77	2.12	2.38	2.58
	df	2.83	5.86	8.44	11.1	13.6
5	d_2	1.19	1.74	2.10	2.36	2.56
	df	4.59	9.31	13.9	18.4	22.6
10	$d_2^;$	1.16	1.72	2.08	2.34	2.55
	df	8.99	18.4	27.6	36.5	44.9
15	$d_2^;$	1.15	1.71	2.07	2.34	2.54
	df	13.4	27.5	41.3	54.6	67.2
20	$d_2^;$	1.14	1.70	2.07	2.33	2.54
	df	17.8	36.5	55.0	72.7	89.6
∞	$d_2^;$	1.13	1.69	2.06	2.33	2.53

Intermediate values for $d_2^;$ and df may be obtained by interpolation, or from the reference from which this table was adapted. Adapted from Lloyd S. Nelson. *J. Qual. Tech.* 7 No. 1. January 1975. ©American Society for Quality Control, Used by permission.

Table A.2. Z-Factors for Confidence Intervals for the Normal Distribution

Confidence Level, %	Z-Factor	
	One-Sided Interval	Two-Sided Interval
50		0.68
67		1.00
75		1.15
90	1.282	1.645
95	1.645	1.960
95.28		2.000
99	2.326	2.575
99.5	2.576	
99.74		3.000
99.9	3.090	
99.9934		4.000
99.99995		5.000
$100 - 10^{-9}$		6.000

Table A.3. Student t Variate

*	80%	90%	95%	98%	99%	99.73% Z=3
**	90%	95%	97.5%	99%	99.5%	99.85%
Df	$t_{.90}$	$t_{.95}$	$t_{.975}$	$t_{.99}$	$t_{.995}$	$t_{.9985}$
1	3.078	6.314	12.706	31.821	63.657	235.80
2	1.886	2.920	4.303	6.965	9.925	19.207
3	1.638	2.353	3.182	4.541	5.841	9.219
4	1.533	2.132	2.776	3.747	4.604	6.620
5	1.476	2.015	2.571	3.365	4.032	5.507
6	1.440	1.943	2.447	3.143	3.707	4.904
7	1.415	1.895	2.365	2.998	3.499	4.530
8	1.397	1.860	2.306	2.896	3.355	4.277
9	1.383	1.833	2.362	2.821	3.250	4.094
10	1.372	1.812	2.228	2.764	3.169	3.975
11	1.363	1.796	2.201	2.718	3.106	3.850
12	1.356	1.782	2.179	2.681	3.055	3.764
13	1.350	1.771	2.160	2.650	3.012	3.694
14	1.345	1.761	2.145	2.624	2.977	3.636
15	1.341	1.753	2.131	2.602	2.947	3.586
16	1.337	1.746	2.120	2.583	2.921	3.544
17	1.333	1.740	2.110	2.567	2.898	3.507
18	1.330	1.734	2.101	2.552	2.878	3.475
19	1.328	1.729	2.093	2.539	2.861	3.447
20	1.325	1.725	2.086	2.528	2.845	3.422
25	1.316	1.708	2.060	2.485	2.787	3.330
30	1.310	1.697	2.042	2.457	2.750	3.270
40	1.303	1.684	2.021	2.423	2.704	3.199
60	1.296	1.671	2.000	2.390	2.660	3.130
∞	1.282	1.645	1.960	2.326	2.576	3.000

*Columns to be used in calculating corresponding two-sided confidence interval. Excerpted from "Experimental Statistics" NBS Handbook 91. Last column from B. J. Joiner, *J. Research* NBS.
**Columns to be used for a corresponding one-sided test.

Table A.4. Factors for Computing Two-Sided Confidence Intervals for σ

Degrees of Freedom	$\alpha = .05$ (95% confidence)		$\alpha = .01$ (99% confidence)		$\alpha = .001$ (99.9% confidence)	
	B_U	B_L	B_U	B_L	B_U	B_L
1	17.79	.358	86.31	.297	844.4	.248
2	4.86	.458	10.70	.388	33.3	.329
3	3.18	.518	5.45	.445	11.6	.382
4	2.57	.559	3.89	.486	6.94	.422
5	2.25	.590	3.18	.518	5.08	.453
6	2.05	.614	2.76	.544	4.13	.478
7	1.92	.634	2.50	.565	3.55	.500
8	1.82	.651	2.31	.583	3.17	.519
9	1.75	.666	2.17	.599	2.89	.535
10	1.69	.678	2.06	.612	2.69	.549
15	1.51	.724	1.76	.663	2.14	.603
20	1.42	.754	1.61	.697	1.89	.640
25	1.36	.775	1.52	.721	1.74	.667
30	1.32	.791	1.46	.740	1.64	.688
35	1.29	.804	1.41	.755	1.58	.705
40	1.27	.815	1.38	.768	1.52	.720
45	1.25	.824	1.35	.779	1.48	.732
50	1.23	.831	1.33	.788	1.45	.743

Excerpted from "Experimental Statistics", NBS Handbook 91.

Table A.5. Factors for Computing Two-Sided Tolerance Intervals for a Normal
                Distribution

n/p	τ = 0.95				τ = 0.99			
	.90	.95	.99	.999	.90	.95	.99	.999
2	32.02	37.67	48.43	60.57	160.19	188.49	242.30	303.05
3	8.38	9.92	12.86	16.21	18.93	22.40	29.06	36.62
4	5.37	6.37	8.30	10.50	9.40	11.15	14.53	18.38
5	4.28	5.08	6.63	8.42	6.61	7.86	10.26	13.02
6	3.71	4.41	5.78	7.34	5.34	6.34	8.30	10.55
7	3.37	4.01	5.25	6.68	4.61	5.49	7.19	9.14
8	3.14	3.73	4.89	6.23	4.15	4.94	6.47	8.23
9	2.97	3.53	4.63	5.90	3.82	4.55	5.97	7.60
10	2.84	3.38	4.43	5.65	3.58	4.26	5.59	7.13
15	2.48	2.95	3.88	4.95	2.94	3.51	4.60	5.88
20	2.31	2.75	3.62	4.61	2.66	3.17	4.16	5.31
25	2.21	2.63	3.46	4.41	2.49	2.97	3.90	4.98

p = proportion of population to be covered (e.g., .90 = 90%).
τ = probability of inclusion (e.g., .95 = 95%).

Excerpted from "Experimental Statistics", NBS Handbook 91.

Table A.6. Critical Values for the F Test, $F_{0.975}$

df_D	df_N									
	1	2	4	6	8	10	15	20	30	40
1	648	800	900	937	957	969	983	993	1001	1006
2	38.5	39.0	39.2	39.3	39.4	39.4	39.4	39.4	39.5	39.5
4	12.2	10.6	9.6	9.2	9.0	8.8	8.7	8.6	8.5	8.4
6	8.8	7.3	6.2	5.8	5.6	5.5	5.3	5.2	5.1	5.0
8	7.6	6.1	5.0	4.6	4.4	4.3	4.1	4.0	3.9	3.8
10	6.9	5.5	4.5	4.1	3.8	3.7	3.5	3.4	3.3	3.3
15	6.2	4.8	3.8	3.4	3.2	3.1	2.9	2.8	2.6	2.6
20	5.9	4.5	3.5	3.1	2.9	2.8	2.6	2.5	2.4	2.3
30	5.6	4.2	3.2	2.9	2.6	2.5	2.3	2.2	2.1	2.0
40	5.4	4.0	3.1	2.7	2.5	2.4	2.2	2.1	1.9	1.9
60	5.3	3.9	3.0	2.6	2.4	2.3	2.1	1.9	1.8	1.7
120	5.2	3.8	2.9	2.5	2.3	2.2	1.9	1.8	1.7	1.6
∞	5.0	3.7	2.8	2.4	2.2	2.1	1.8	1.7	1.6	1.5

For use for a one-tailed test of equality of standard deviation estimates at 2.5% level of confidence, or for a two-tailed test at 5% level of confidence. df_N and df_D refer to degrees of freedom of variances of numerator and denominator of F test, respectively. Excerpted from "Experimental Statistics", NBS Handbook 91.

Table A.7. Values for Use in the Dixon Test for Outliers

Statistic	Number of Observations, n	Risk of False Rejection			
		0.5%	1%	5%	10%
	3	.994	.988	.941	.886
	4	.926	.889	.765	.679
τ_{10}	5	.821	.780	.642	.557
	6	.740	.698	.560	.482
	7	.680	.637	.507	.434
	8	.725	.683	.554	.479
τ_{11}	9	.677	.635	.512	.441
	10	.639	.597	.477	.409
	11	.713	.679	.576	.517
τ_{21}	12	.675	.642	.546	.490
	13	.649	.615	.521	.467
	14	.674	.641	.546	.492
	15	.647	.616	.525	.472
	16	.624	.595	.507	.454
τ_{22}	17	.605	.577	.490	.438
	18	.589	.561	.475	.424
	19	.575	.547	.462	.412
	20	.562	.535	.450	.401

Tabulated values obtained from reference NBS Handbook 91. Original reference: W. J. Dixon, "Processing Data Outliers", Biometrics BIOMA, March, 1953, Vol. 9, No. 1, pp 74–89.

Table A.8. Values for Use in the Grubbs Test for Outliers

Number of Data Points	Risk of False Rejection				
	0.1%	0.5%	1%	5%	10%
	Critical Values				
3	1.155	1.155	1.155	1.153	1.148
4	1.496	1.496	1.492	1.463	1.425
5	1.780	1.764	1.749	1.672	1.602
6	2.011	1.973	1.944	1.822	1.729
7	2.201	2.139	2.097	1.938	1.828
8	2.358	2.274	2.221	2.032	1.909
9	2.492	2.387	2.323	2.110	1.977
10	2.606	2.482	2.410	2.176	2.036
15	2.997	2.806	2.705	2.409	2.247
20	3.230	3.001	2.884	2.557	2.385
25	3.389	3.135	3.009	2.663	2.486
50	3.789	3.483	3.336	2.956	2.768
100	4.084	3.754	3.600	3.207	3.017

Tabulated values obtained in part from ASTM E-178 which should be consulted for more extensive tables. Original reference: F. E. Grubbs and G. Beck, "Extension of Sample Sizes and Percentage Points for Significance Tests of Outlying Observations", Technometrics, TCMTA, Vol. 14, No. 4, Nov. 1972, pp 847–54.

Table A.9. Values for Use in the Youden Test to Identify Consistent Outlying
Performance Approximate 5% Two-Tail Limits for Ranking Scores*

Number of Parti-cipants	3	4	5	6	7	8	9	10	11	12	13	14	15
					Number of Materials								
3		4	5	7	8	10	12	13	15	17	19	20	22
		12	15	17	20	22	24	27	29	31	33	36	38
4		4	6	8	10	12	14	16	18	20	22	24	26
		16	19	22	25	28	31	34	37	40	43	46	49
5		5	7	9	11	13	16	18	21	23	26	28	31
		19	23	27	31	35	38	42	45	49	52	56	59
6	3	5	7	10	12	15	18	21	23	26	29	32	35
	18	23	28	32	37	41	45	49	54	58	62	66	70
7	3	5	8	11	14	17	20	23	26	29	32	36	39
	21	27	32	37	42	47	52	57	62	67	72	76	81
8	3	6	9	12	15	18	22	25	29	32	36	39	43
	24	30	36	42	48	54	59	65	70	76	81	87	92
9	3	6	9	13	16	20	24	27	31	35	39	43	47
	27	34	41	47	54	60	66	73	79	85	91	97	103
10	4	7	10	14	17	21	26	30	34	38	43	47	51
	29	37	45	52	60	67	73	80	87	94	100	107	114
11	4	7	11	15	19	23	27	32	36	41	46	51	55
	32	41	49	57	65	73	81	88	96	103	110	117	125
12	4	7	11	15	20	24	29	34	39	44	49	54	59
	35	45	54	63	71	80	88	96	104	112	120	128	136
13	4	8	12	16	21	26	31	36	42	47	52	58	63
	38	48	58	68	77	86	95	104	112	121	130	138	147
14	4	8	12	17	22	27	33	38	44	50	56	61	67
	41	52	63	73	83	93	102	112	121	130	139	149	158
15	4	8	13	18	23	29	35	41	47	53	59	65	71
	44	56	67	78	89	99	109	119	129	139	149	159	169

*Approximate, because of rounding. From W.J. Youden. p. 149 of NBS Special Publication 300, Vol. 1

Table A.10. Values for Use in the Cochran Test for Extreme Values for Variance

Numbers of Variances Compared	(5% Risk of Wrong Decision) Number of Replicate Values Used to Compute Each Variance							
	2	3	4	5	6	7	10	∞
2	.9985	.9750	.9392	.9057	.8772	.8534	.8010	.5000
3	.9969	.8709	.7977	.7457	.7071	.6771	.6167	.3333
4	.9065	.7679	.6841	.6287	.5895	.5598	.5017	.2500
5	.8412	.6838	.5981	.5441	.5065	.4783	.4214	.2000
6	.7808	.6161	.5321	.4803	.4447	.4184	.3682	.1667
7	.7271	.5612	.4800	.4307	.3974	.3726	.3259	.1429
10	.6020	.4450	.3733	.3311	.3029	.2823	.2439	.1000
20	.3894	.2705	.2205	.1921	.1735	.1602	.1357	.0500
30	.2929	.1980	.1593	.1377	.1237	.1137	.0958	.0333
40	.2370	.1576	.1259	.1082	.0968	.0887	.0745	.0250
60	.1737	.1131	.0895	.0765	.0682	.0623	.0520	.0167

Extract from Table 14.1, Eisenhart, Hastay, and Wallis, Selected Techniques of Statistical Analysis, McGraw-Hill Book Co. (1947). *Each variance must be estimated with the same number of degrees of freedom. Calculate s^2 (largest)/$\Sigma\; s_i^2$. If ratio exceeds tabulated value, assume largest value is extreme with 95% confidence (5% risk of wrong decision).*

Table A.11. A Short Table of Random Numbers

84	81	91	06	12	11	83	06	10	34	23	29	00	64	02	40
03	45	86	26	38	48	95	32	05	35	84	07	39	35	10	38
07	28	77	46	12	64	45	16	83	53	60	92	20	91	90	97
63	45	60	89	58	63	83	56	80	54	84	46	74	23	60	65
43	11	37	49	95	08	49	10	47	94	92	39	30	03	47	02
76	15	84	79	23	21	86	11	60	28	03	52	46	32	06	27
04	03	08	25	83	32	44	94	53	88	60	08	76	82	04	71
89	56	57	39	95	37	24	65	22	14	98	17	94	98	65	42
80	05	61	14	67	01	34	06	71	09	36	06	84	50	53	36
71	44	13	03	43	65	06	70	39	80	65	33	78	79	75	51
52	04	65	03	16	07	68	14	97	17	57	59	93	81	08	65
02	55	02	62	36	92	40	47	28	54	55	71	82	29	32	85
04	55	10	83	50	73	39	80	84	60	99	95	88	99	98	50
95	86	97	72	93	74	72	26	30	96	84	01	66	00	72	83
89	39	22	82	05	02	30	84	98	92	46	47	90	34	80	01
85	58	68	28	86	31	34	54	59	97	03	90	48	98	90	09
99	31	09	47	96	81	26	06	17	29	63	46	27	65	01	13
76	07	69	89	27	70	76	08	23	06	17	73	02	67	69	38
01	86	37	92	06	46	66	85	16	60	52	69	77	67	72	66
36	18	34	99	29	68	96	05	59	11	90	89	66	17	28	02
97	83	05	39	60	67	45	54	55	01	60	70	95	72	59	94
51	20	31	58	93	26	48	84	83	38	84	39	93	00	86	81
12	20	88	06	82	14	18	97	61	44	93	75	32	18	79	67
76	14	61	54	07	65	87	50	00	37	61	42	85	03	65	06
93	54	70	44	15	07	55	49	29	97	58	37	13	04	72	99
63	69	70	66	12	85	99	31	59	41	87	05	62	02	44	74
31	10	56	98	92	38	71	51	79	93	38	67	28	26	49	01
03	17	76	87	23	64	36	89	01	51	75	96	47	78	71	53
98	17	65	89	52	19	13	94	63	62	57	47	87	79	17	02
49	00	59	40	56	10	94	77	60	24	27	11	93	91	51	24
21	83	49	05	47	29	67	88	46	94	41	10	17	85	02	36
62	92	13	02	64	33	12	83	29	61	29	59	57	70	64	80
13	73	76	47	12	42	74	66	91	82	86	71	66	76	50	39
50	09	97	21	15	79	18	13	49	76	59	54	31	55	17	36
35	58	59	60	27	43	71	03	89	74	42	96	77	19	50	22
70	01	37	01	96	64	76	74	10	53	86	33	89	46	58	91
58	45	68	43	11	05	26	58	64	72	79	91	54	78	88	26
69	30	79	36	18	47	91	25	34	54	77	24	62	13	60	88
28	89	94	85	90	09	13	55	19	24	95	26	76	57	45	33
50	26	02	28	11	39	48	44	17	23	69	94	84	23	22	49
64	53	30	65	99	05	32	13	46	47	35	54	68	46	31	41
57	29	23	22	8	20	27	04	09	65	45	56	58	68	24	75
56	15	11	92	54	06	32	09	23	72	02	90	80	86	61	79

APPENDIX **B**

Glossary

The following terms are defined in the context in which they are used in this book. In general, they agree in essence with the usage of most authors in this field. The definitions come from various sources, including the author's synthesis of current usage.

There are many statistical terms that are not used in the present book but which may be encountered by the reader who may wish to delve into statistics more deeply. The author highly recommends the books by Freund and Williams [1] and by Kendall and Buckland [2] for definitions of extensive lists of statistical terms. The former book also contains formulas for many statistical calculations.

For those who want more detailed information on statistical terms and concepts, the encyclopedia edited by Katz, Johnson, and Read is highly recommended [3].

Accuracy – The accuracy of a measurement process refers to, and is determined by the degree of conformity to the truth, i.e., the true value.

Analysis of variance (ANOVA) – A statistical procedure designed to evaluate the variance arising from one or more sources in experimental data.

Assignable cause – In quality assurance studies, the term represents excess variability (sometimes bias as well) the source(s) of which can be identified and possibly corrected or minimized by appropriate actions. (Also called special cause). See also chance cause.

Attribute – A qualitative characteristic of a population or individuals belonging to a population.

Attribute data – Data pertaining to some selected attribute or attributes of a population. An example is the number of defects of a particular kind observed in a given sample of a population.

Average deviation – The arithmetic mean of the absolute values of the deviations of the n individuals, X_i, in a data set from the mean, $\overline{X}$, computed from them. It is computed from

$$\text{A.D.} = [\Sigma \,|\, X_i - \overline{X} \,|\,] / n$$

Bias – A systematic error which is inherent in a method
by some artifact or idiosyncracy of the measurement sy
positive and negative and several kinds can exist concurre
can be evaluated in many cases. Bias is measured as th
limiting mean and the true value measured by a system tha   ...trol.
(Also called offset by some metrologists.)

Boxplot – A graphical way to represent the median, quartiles and outliers in a
dataset by means of a box, the ends of which represent the quartiles and the middle
the median. Outliers are denoted by asterisks.

Censoring – A characteristic of survival data, in which the time of an event is the
outcome of interest. If the event of interest occurs, the time is recorded and the data
is considered uncensored. Should the event not occur, the last time observed is
recorded and the event time is considered censored.

Chance cause – In quality assurance terminology, a source of variation that is always
present in a measurement system and which is responsible for the standard deviation
observed in a system in statistical control. Metrologists believe that a number of
chance causes can be at play in well controlled measurement systems due to lack of
understanding of their sources and/or inability to control them to desired tolerances.
Ordinarily, they are accepted as being tolerable. Improvement in the accuracy of
measurement systems often results from identification of components of chance
causes and taking permanent corrective actions to control them. (Also called common causes).

Chemometrics – A new discipline involving statistics, mathematics, and chemistry,
that deals with the interrelations of these sciences.

Coefficient of variation – The ratio of the estimated standard deviation divided by
the average value measured. It is calculated by

$$CV = s / \overline{X}$$

Confidence interval – The range of values within which some parameter may be
expected to lie with a stated confidence. The confidence level is stated as a percentage which is related to the probability of inclusion of the parameter of interest.

Correlation coefficient – A number that serves to indicate the degree of correlation
between two quantitative variables. It is calculated by

$$\rho = \frac{\sigma_{xy}}{\sigma_x \sigma_y}$$

and is estimated by

$$r = \frac{\Sigma (X_i - \overline{X})(Y_i - \overline{Y})}{\sqrt{[\Sigma(X_i - \overline{X})^2 \; \Sigma(Y_i - \overline{Y})^2]}}$$

where both ρ and r lie between -1 and $+1$. A value of $+1$ indicates perfect positive correlation (Y increases with X) while -1 indicates perfect negative correlation (Y decreases with X). A value of zero indicates no correlation.

Critical value approach – A technique used in statistical hypothesis testing where the test statistic value is compared to a table value (called the critical value) based on a certain distribution and level of significance (and possibly sample size). The null hypothesis is rejected if the test statistic value exceeds the table (critical) value.

Data points – A value for some variable that may be used as an entity for calculation or statistical analysis. It may be a single value or the mean of several individual values.

Degrees of freedom – The excess of the number of variables over the number of statistical parameters to be determined.

Dependent variable – A parameter (usually designated the Y value) whose value is determined by a functionally related independent variable.

Design of experiments – A branch of chemometrics that involves statistically based plans to minimize effort and maximize experimental information.

Deviation – The numerical difference, d, of an observed value from a computed value or a true value. The computed value may be the mean of a data set or that obtained from an equation, expressing or predicting the values of a dependent variable associated with independent variables. The use will make it clear what kind of deviation is indicated. Deviations are computed by the equation

$$d = X_o - X_c$$

A positive sign indicates that the observed value is larger than the computed or true one, and conversely.

Distribution – A statistical function associating probability with data points.

Error – The difference between a measured value and the true value of the parameter measured. It is due to random error and systematic error.

Eye estimate – The result of an estimation process in which the only instrument used is the eye. The term is used especially to denote a line fitted to plotted data points in which the best fit is judged by the eye.

Finite population – A well defined set consisting of a finite number of elements. Withdrawal of any member without replacement distorts the population.

Gaussian distribution – A theoretical frequency distribution described by an expression developed by Karl Friedrich Gauss, a German mathematician (1855), which is

$$y = \frac{1}{\sigma\sqrt{2\pi}} e^{-(x-\mu)^2/2\sigma^2}$$

where μ is the mean and σ is the standard deviation, both of the population.

Harmonic mean – The reciprocal of the arithmetic mean of reciprocals of two or more quantities. It is especially useful for considering ratios of things as data, such as rates, for example. It is calculated by the expression

$$H = \frac{n}{\sum \frac{1}{x}}$$

where n is the number of quantities averaged. Thus the harmonic mean cost of apples costing 40 cts/kg and 50 cts/kg is 44.4 cts/kg.

Histogram – A graphical way to represent a frequency distribution of a population or sample by means of bars, the width of which represents the class intervals and the height the corresponding frequency of occurrence of each class.

Independent variable – A parameter (usually the X value) whose values can be chosen at will to determine the value of a functionally related dependent variable.

Individuals – Conceivable constituent parts of a population.

Infinite population – A population, well defined by characteristics, that is limitless in number and unaffected by withdrawal of any present or past withdrawal of a finite sample.

Kaplan Meier survival estimate – A method of calculating an estimate of the survival function in survival data analysis. The survival function estimate is calculated at each event time (censored or uncensored) using the number of units at risk during the time interval. Also called the Product Limit Survival Estimate.

Limiting mean – The value that the mean of a stable process approaches as the number of measurements is indefinitely increased.

Log rank test – A statistical test used in survival data analysis for determining whether the survival curves of two or more groups of units significantly differ.

Macro – A technique used in statistical programming where computer code is written to perform a specific calculation. The code is usually saved for future use and then invoked with a command.

Median – The middle value (or average of the two middle values) of a set of observations when the individual values are arranged in ascending order of the numerical size.

Method of averages – A mathematical procedure to adjust the parameters of a function so that the algebraic sum of the residuals is zero.

Method of least squares – A mathematical procedure to adjust the parameters of a function so that the sum of the squares of the deviations of the observed quantities from the respective calculated values is a minimum.

MINITAB – A statistical software package providing capabilities for the analysis and graphical presentation of data.

Normal distribution – A frequency distribution that is bell-shaped, symmetrical, and of infinite extent. It is mathematically described by an expression due to Karl Friedrich Gauss and often called the Gaussian distribution.

Normal probability plot – A graphical way to judge whether a dataset exhibits departures from the normality assumption. A plot of the data value versus probability plot position should yield an approximately straight line if there are no extreme departures from normality.

Observation – The value noted of some variable or instrumental reading as the result of experimentation or of a natural occurrence. Ordinarily, this is a single data point.

OC curve – Operating characteristic curve. A graphical depiction of the discriminatory power of a particular statistical test.

Offset – See bias.

Outlier – A value in a data set which appears to deviate markedly from other members of the same sample in which it occurs, and has low statistical probability of belonging to the same population.

P value approach – A technique used in statistical hypothesis testing where the probability of the test statistic or a more extreme value is calculated under the assumption that the null hypothesis is true. Generally, if a p value is less than 0.05 it is thought that the chance of the null hypothesis holding true is so unlikely that the null hypothesis is rejected in favor of the alternative hypothesis.

Poisson distribution – A model with a discrete distribution used to describe random phenomena in which the probability of occurrence is small but constant.

Population – A generic term denoting any finite or infinite collection of individual things, objects, or events: in the broadest concept, an aggregate determined by some property that distinguishes things that do and do not belong. The term, universe, is used interchangeably with population.

Power – A statistical term used to represent the probability a statistical test has of detecting a given difference in a dataset, when in reality that difference exists.

Precision – The degree of mutual agreement characteristic of independent measurements of a single quantity yielded by repeated applications of the process under specified conditions.

Probable error – A once popular way to express random error, but little used today. It is the value that defines the 50% portion of the error distribution in a normal distribution. Half the errors will be smaller and half larger than the probable error. It is calculated by P.E. $= 0.6747 \sqrt{[\Sigma d^2 /(n-1)]}$. The probable error is thus 0.6747σ or P.E. $\approx 2/3\sigma$.

Probability – The likelihood of the occurrence of any particular form of any event, estimated as the ratio of number ways or times that it could occur in that form to the number of ways that it could occur in any form.

Product limit survival estimate – A method of calculating an estimate of the survival function in survival data analysis. The survival function estimate is calculated at each event time (censored or uncensored) using the number of units at risk during the time interval. Also called the Kaplan Meier survival estimate.

Propagation of error – How the individual errors of measurement are propagated in a final computed result.

Proportional hazards – A key assumption for many types of multi-group survival data analysis. The hazard, or instantaneous risk of failure, in one group, is a constant multiple of that in another group under the proportional hazards assumption. It can be shown mathematically that proportional hazards implies that the survival functions for the groups do not cross when graphically displayed.

Quaesitum – The actual value of a quantity, that is to say its true value.

Random error – Error due to variability causes. The net random error of a measurement process can cause either positive or negative errors, both small and large within limits. Random errors will average out to zero for a large number of independent measurements.

Random sample – A sample drawn from a population, using a randomization process.

Range – The difference between the largest and the smallest values in a data set. Ordinarily, it is the absolute value and does not take into account the time sequence of the respective observations. In the case of duplicates, the order of observation may be considered in special cases, i.e., the difference of the second minus the first may be recorded to determine whether there is any significant difference due to order of measurement.

Relative standard deviation – The coefficient of variation expressed as a percentage. Thus,

$$RSD = 100(s / \overline{X})$$

Repeatability – The standard deviation of measurements made in close succession (i.e., the short term standard deviation). In ISO terminology, it represents $2 \times \sqrt{2}$ times this standard deviation which corresponds to the expected difference of two successive observations with 95% confidence. Some US methods quote the repeatibility as 2 × this standard deviation.

Reproducibility – The standard deviation of observations made over a protracted period of time (the long term standard deviation). In ISO terminology, it is $2 \times \sqrt{2}$ times this standard deviation. Some US methods quote the reproducibility as 2 × this standard deviation.

Ruggedness test – An experimental design described by W.J. Youden to verify that assumed tolerances are adequate for control of a measurement process. The tolerances for n variables can be tested using n + 1 measurements. Ruggedness tests are special cases of fractional factorial designs.

Sample – A portion of a population. It may consist of an individual or groups of individuals. It may refer to objects, materials, or measurements, conceivable as part of a larger group that could have been considered.

Sample size – A numerical calculation designed to determine, with a specific type I error rate and power, how many observations are required to detect a specified difference assuming a defined amount of variability in the data.

Simulation – A technique used in conjunction with statistical macro programming where the same calculation is performed hundreds or thousands of times to see the large sample properties of a given statistic or hypothesis.

Standard deviation – The positive square root of the variance. It is calculated from the expression.

$$S.D. = \sqrt{[\sum(X_i - \overline{X})^2 /(n-1)]}$$

So calculated, it represents the standard deviation of individuals about the mean. See text for distinction between σ and s.

Standard deviation of the mean – Also called the standard error of the mean. It represents the standard deviation of means of size, n, about the grand mean, i.e., the grand average of all means of that size. It is calculated by

$$S.D. \text{ of mean} = (\sigma \text{ or } s)/\sqrt{n}$$

Standard error of the mean – See standard deviation of the mean.

Survival data – A type of data encountered in which the time of a specific event is of interest. Under Type I censoring, units are observed for a set period of time. If the event of interest occurs during this time, the time is recorded and the data is considered uncensored. Should the event not occur, the last time observed is recorded and the event time is considered censored. Under Type II censoring, units are observed until a set number of failures occur. If the event of interest occurs during this time, the time is recorded and the data is considered uncensored. Should the event not occur, the last time observed is recorded and the event time is considered censored.

Systematic error – An error of the same size and sign that occurs due to a constant recurring cause. It will produce a bias or offset of the result of a measurement process. Systematic errors never average out and the magnitude is independent of the number of repetitions of the measurement process.

Tolerance interval (engineering) – The engineering tolerance interval represents limits of acceptability of some characteristic of a population, usually prescribed by a design engineer.

Tolerance interval (statistical) – The statistical tolerance interval represents limits within which a stated percentage of a population is expected to lie, based on the statistical variability of some characteristic of a population.

Trueness – The closeness of agreement between the average value obtained from a large series of test results and an accepted reference value. The measure of trueness is usually expressed in terms of bias.

Type I error – A statistical term used to represent the probability a statistical test has of detecting a given difference in a dataset, when in reality that difference does not exist.

Type II error – A statistical term used to represent the probability a statistical test has of not detecting a given difference in a dataset, when in reality that difference exists.

Uncertainty – The range of values within which the true value is estimated to lie. It is the best estimate of the possible inaccuracy due to both random and systematic error.

Universe – A term used interchangeably with Population.

Variable data – Data obtained as the result of measurements of a variable.

Variance – The value approached by the average of the sum of the squares of deviations of individual measurements from the mean. Mathematically, it may be expressed as

$$V = \sigma^2 = \Sigma(X_i - m)^2 / n \ \text{ as n} \rightarrow \infty$$

Ordinarily, it cannot be known, but only its estimate, s^2, which is calculated by

$$s^2 = \Sigma(X_i - \overline{X})^2 / (n-1)$$

Weighting – The weight of an observation is its relative value among others of the same set. It represents the relative confidence in that observation in comparison with other observations of the same quantity. In the context of the definition, an observation may be an individual value or the mean of a set of values.

Youden plot – A graphical presentation of data, recommended first by W.J. Youden, in which the result(s) obtained by a laboratory on one sample are plotted respect to the result(s) it obtained on a similar sample. Depending on the relation of the plotted point to the "true value", it can be decided whether discrepant results are due to bias, imprecision, or both causes.

REFERENCES

[1]   Freund, J.E. and F.J. Williams, "Dictionary/Outline of Basic Statistics", McGraw-Hill Book Co., New York (1966).

[2]   Kendall, M.G. and W.R. Buckland, "A Dictionary of Statistical Terms", 3rd Edition, Hafner Publishing Co., New York (1971).

[3]   Katz, S., N.L. Johnson, and C.B. Read, Eds. "Encyclopedia of Statistical Sciences" (9 volumes), J. Wiley & Sons, New York (1982).

APPENDIX $\mathbf{C}$

Answers to Selected Numerical Exercises [*]

CHAPTER 3

E 3-4.

X	20.20	20.35	20.37	20.45	20.50	20.50	20.60	20.65	20.65	20.70
i	1	2	3	4	5	6	7	8	9	10
F_i	5	15	25	35	45	55	65	75	85	95

The plot on arithmetic probability paper is sufficently well represented by a straight line to justify application of normal statistics.

Given

$$\overline{X} = 20.497 \quad s = 0.159 \quad g_1 = -0.371 \quad g_2 = 1.213$$

E 3-5.

X	5.9	6.9	7.4	8.0	8.4	9.0	9.6	10.6	13.0	16.5
i	1	2	3	4	5	6	7	8	9	10

[*] Readers may evaluate their responses to essay questions by reference to the text. The index may be consulted to provide guidance where related discussions may be found.

F_i	5	15	25	35	45	55	65	75	85	95

A plot on arithmetic probability paper consists of a curve that cannot be fitted to a straight line. When plotted on logarithmitic probability paper, a reasonable straight line fit is possible. The data appear to be lognormally distributed.

CHAPTER 4

E 4-3.

$$\overline{X} = 51.0 \quad s = 2.6 \quad df = 9 \quad cv = 0.051 \quad RSD = 5.1\%$$

E 4-4.

$$\overline{X} = 151.0 \quad s = 2.6 \quad df = 9 \quad cv = 0.017 \quad RSD = 1.7\%$$

The values for the data for E 4-4 differ from DE 4-3 by 100, but the absolute differences from the means are exactly the same. Hence the value for s is the same in each case. Thus, if one were measuring a dimension of either 51 or 151 cm using a tape, the standard deviation could be the same in each case if the variability were due entirely to that of estimation of the final reading. The values for cv and RSD will be smaller in the second case because $\overline{X}$ is larger.

E 4-5.

Set No.	Measured Values	r	s	s^2	df
1	20.5, 21.0, 20.6	0.5	.265	.0697	2
2	20.9, 20.7, 20.9	0.2	.115	.0132	2
3	20.7, 21.3, 20.9	0.6	.306	.0936	2
4	20.8, 20.8, 20.5	0.3	.173	.0299	2
5	21.0, 21.1, 20.6	0.5	.264	.0697	2

$$\overline{R} = 0.42 \quad d_2^* = 1.74 \quad df = 9.3\,i/e.9 \quad s = .42/1.74 = 0.24$$

$$s_p = \sqrt{\frac{(0697\times2+.0132\times2+.0936\times2+0299\times2+.0697\times2)}{2+2+2+2+2}}$$

$$s_p = 0.23 \quad df = 10$$

E 4-6.

$$s_T = \sqrt{(1.10+2.96+0.26+0.72+6.25)}$$

$$s_T = 3.36$$

E 4-7.

Set No.	X_f	X_s	R	d	d^2
1	20.7	21.0	0.3	0.3	0.090
2	21.1	20.9	0.2	0.2	0.040
3	21.0	21.1	0.1	0.1	0.010
4	23.0	23.0	0.0	0.0	0.000
5	18.3	17.8	0.5	0.5	0.250
6	19.9	20.0	0.1	0.1	0.010

$$\overline{R} = 0.20 \quad d_2^* = 1.18 \quad df = 5 \quad s = 0.20/1.18 = 0.17$$
$$s = \sqrt{[(.09+.04+.01+0+.25+.01)/12]}$$
$$s = 0.18 \, df = 6$$

E 4-9.

Occasion	s	n	s^2	df
1	0.75	10	0.562	9
2	1.45	5	2.102	4
3	1.06	7	1.124	6
4	2.00	7	4.000	6
5	1.25	12	1.562	11

$$s_p = \sqrt{\frac{0.562\times9+2.102\times4+1.124\times6+4.000\times6+1.562\times11}{9+4+6+6+11}}$$
$$s_p = 1.31 \quad df = 36$$

E 4-10.

The standard deviation of a mean represents the dispersion of repetitive determinations of means, each based on the same number of replicates. One can either make a
sufficient number of replicate measurements and calculate their standard deviation
using the formula for a set of measurements given Chapter 4.2, or by estimating the
standard deviation of single measurements and calculating it, using the appropriate
equation.

s	n	s^2	$s_{\bar{x}}$
1.22	10	(1.49)	(.39)
(1.86)	(4)	3.45	0.929
(1.40)	7	(1.96)	0.529
2.44	(21)	(5.95)	0.532
1.55	—	(2.40)	—

CHAPTER 5

E 5-1.

189.8	limited by the value 12.3
0.8713	limited by 14.32 and 7.575
0.0000285	limited by the value 0.0000285
10692	limited by 10670

12.4
127
.0326
12.4

E 5-3.

You might say that the measurement data leads to the conclusion that there is 95% confidence that 95% of the samples in the lot would have compositions within the range of 18.5 to 22.1.

E 5-4.

a. The plot leads to the conclusion that the data appear to be normally distributed and that normal statistics can be used.
b. 95% C.I.

$$\overline{X} = 3.557 \quad s = 0.046 \quad n = 9 \quad df = 8 \quad t = 2.306$$

$$\overline{X} = 3.557 \pm 2.306 \times .046/\sqrt{9}$$
$$\overline{X} = 3.557 \pm 0.035$$

c. 95,95 T.I.

$$3.557 \pm 3.53 \times .046 = 3.56 \pm 0.16$$

95,99 T.I.

$$3.557 \pm 4.63 \times .046 = 3.56 \pm 0.21$$

d.

$$s_T^2 = s_a^2 + s_s^2$$
$$.046^2 = .03^2 + s_s^2$$
$$s_s = .035$$

95% C.I. mean $= 3.557 \pm 2.042 \times .03/\sqrt{9} = 3.557 \pm 0.020$

95,95 T.I. $= 3.557 \pm 0.020 \pm 3.53 \times 0.035 = 3.56 \pm 0.14$

E 5-5.

Analysts data $\overline{X} = 14.534$, s $= 0.16$, and df $= 6$; for 95% confidence, $B_L = .614$ and $B_U = 2.05$.

$$95 \text{ C.I. for s} = .614 \times .16 \text{ to } 2.05 \times .16 = 0.098 \text{ to } 0.33$$

Since 0.12 is within the above interval, there is no reason to believe at the selected level of confidence of the test that the attained precision differs from the expected precision.

E 5-6.

Lab A results $X_A = 66.21$ $s_A = 0.19$ $n_A = 4$
Lab B results $X_B = 66.61$ $s_B = 0.17$ $n_B = 4$

$$\overline{X}_B - \overline{X}_A = 0.40$$

Use 95% confidence level for the test using Section 3.6 Case II.

$$f = 6 \quad U = 0.31$$

Since 0.40 > 0.31, conclude with 95% confidence that the results differ.

E 5-7.

Analyst's results $X = 14.626$ $s = 0.068$ df $= 6$

a.   Report s $= .068$, based on df $= 6$

b.   95% C.I. $= 14.626 \pm 2.447 \times .068 / \sqrt{7} = 14.626 \pm .062$
      or 14.564 to 14.688
      Ref. Mat. 14.55 to 14.59

Since the reference material value lies within limits of measurement, conclude with 95% confidence that there is no reason to believe that the method produces biased results.

E 5-8.

Choose 95% level of confidence for the F test

$$s_A = 1.75 \quad s^2_A = 3.06 \quad df = 10$$

$$s_B = 3.50 \quad s^2_B = 12.25 \quad df = 6$$

$$F = \frac{12.25}{3.06} = 4.0$$

$$F_c = 4.1$$

Conclude with 95% confidence that the two estimates do not differ

E 5-9.

Complete the following table, filling in all blanks, as possible:

$\overline{X}$	n	df	s_X	$s_{\overline{X}}$	S^2_x	$t_{.975}$	$k_{95,95}$	95%CI	95,95TI
2	(5)	(4)	(.41)	(.18)	(.17)	(2.776)	5.08	.508	(2.1)
4	2	(1)	(.05)	(.035)	(.0025)	(12.706)	(37.7)	.45	(1.9)
6	(8)	7	.20	(.071)	(.040)	(2.365)	(3.73)	(.17)	(.75)
8	6	(5)	(.25)	.102	(.062)	(2.571)	(4.41)	(.26)	(1.10)
10	(10)	(9)	.50	.158	(.25)	(2.262)	(3.38)	(.36)	(1.69)
15	20	(19)	.30	(.067)	(.090)	(2.093)	(2.75)	(.14)	(.82)
20	(4)	(3)	(.87)	(.435)	(.76)	3.182	(6.37)	1.38	(5.5)

CHAPTER 6

E 6-4.

a.   Ranked data set

14.5, 14.6, 14.7, 14.8, 14.9, 14.9, 15.3

The value 15.3 is possibly an outlier

Dixon Test   For 7 data points, calculate r_{10}

$$r_{10} = \frac{15.3 - 14.9}{15.3 - 14.5} = 0.50$$

From Table $r_{10} = 0.507$ at 95% confidence (5% risk)
Conclusion: retain the data point

Grubbs Test $\overline{X} = 14.814$ $s = .261$

$$T = \frac{15.3 - 14.814}{.261} = 1.86$$

T from table = 1.938 for 5% risk (95% confidence)

Conclusion: retain the data point

b.   Ranked data set

7.1, 8.0, 8.0, 8.2, 8.3, 8.3, 8.4, 8.5, 8.7, 8.9

7.1 is suspected to be an outlier

Dixon Test   Calculate r_{11}

$$r_{11} = \frac{8.0 - 7.1}{8.7 - 7.1} = .562$$

$$r_{11} = .477 \, (5\% \text{ risk})$$

Conclusion: Rejection is justified

Grubbs Test

$$\overline{X} = 8.24 \quad s = 0.490$$

$$T = \frac{8.24 - 7.1}{.490} = 2.33$$

T from Table A.8 = 2.176 (5% risk)

Conclusion: reject data point

E 6-5.

a.   Ranked $\overline{X}$ values .900, .905, .917, .926, .957
      Check .957
      Dixon Test, Calculate r_{10}

$$r_{10} = \frac{.957 - .926}{.957 - .900} = 0.54 \quad \text{vs. Table value} = .642$$

Conclusion: not an outlier

Grubbs Test $\overline{X} = .921$ $s = .023$ $n = 5$

$$T = \frac{.957 - .921}{.023} = 1.57 \text{ vs. Table value} = 1.67$$

Conclusion: not an outlier

b.   Cochran Test

Ranked s values	.049,	.050,	.053,	.056,	.062
s^2	.00240	.00250	.00281	.00314	.00384

Check .00384 by Cochran's Test

$$\frac{.00384}{.01469} = .261 \text{ vs Table value} = .6838$$

Conclusion: not an outlier

c.

$$s_p = \sqrt{\frac{.0024 \times 2 + .00250 \times 2 + .00281 \times 2 + .00314 \times 2 + .00384 \times 2}{2 + 2 + 2 + 2 + 2}}$$

$$s_p = \sqrt{\frac{.02938}{10}} = .054 \quad df = 10$$

d. $\overline{X} = .921$

e.   Use s value calculated in a.

$$95\% \text{ C.I.} = .921 \pm (2.776 \times .023) / \sqrt{5}$$
$$= .921 \pm .029 \text{ vs. } .95$$

conclusion, not biased at 95% confidence level of test

E 6-6.

$\overline{X}$	n	s	V_i	W^i
10.35	10	1.20	.144	6.94
12.00	5	2.00	.800	1.25
11.10	7	1.50	.321	3.12

$$\overline{\overline{X}} = \frac{10.35 \times 6.94 + 12.00 \times 1.25 + 11.10 \times 3.12}{6.94 + 1.25 + 3.12} = 10.74$$

E 6-7.

Scores

Lab number	Sample Number					Cumulative score
	1	2	3	4	5	
1	9	6	6	4	3	28
2	1	1	3	2	10	17
3	4	8	8	6	8	34
4	5	5	5	8	4	27
5	6	2	7	7	6	28
6	3	3	1	1	1	9
7	10	9	10	9	5	43
8	7	4	9	3	9	32
9	8	10	2	10	7	37
10	2	7	4	5	2	20

The 95% range of test scores from Table 9 is 10 to 45. Conclusion: laboratory 6 produced relatively high scores, consistently. There was no consistent low laboratory.

E 6-8.

The following is an example of one of the many schemes that could be developed.
Column and row selections by colleagues: C = 12, R = 23. Proceed from this point in reverse order of reading, i.e., left-to right, bottom-to-top.

Starting Number = 75

(75),(93),44,(61),(97),18,14,(82),06,(88),20,12,(81),(86),(00),
(93),39,(84),38,(83),(84),48,26

The samples to be analyzed are

06, 12, 14, 18, 20, 26, 38, 39, 44, 48

E 6-9.

Sample No.	06	12	14	18	20	26	38	39	44	48
Arbitrary No.	01	02	03	04	05	06	07	08	09	10

Assign an arbitrary number to facilitate order selection.

Analytical Order	1	6	5	9	8	4	7	10	3	2

Random selection of order. Start at column 09, row 13, starting number = 84. Procede from that point, reading downward.
Order in which arbitrary numbers are found: 01,10,09,06,03,02,07 05,04,08. Use the sequence in which these were found as the analytical order and insert numbers in the table above.

CHAPTER 7

E 7-1.
Using the selected points

X   1 , 6
Y   10 , 62

$$Y = -0.40 + 10.40\ X$$

Other selections would give different equations.

E 7-2.
   Using Σ equations $1 + 2 + 3$ and $4 + 5 + 6$,

$$Y = -1.79 + 10.55\ X$$

Other combinations would give different equations.

E 7-3.

$$Y = -0.53 + 10.20\ X$$

E 7-4.

$$s_a = 1.87 \quad s_b = 0.48$$

E 7-5.
   Equation for original data is

$$Y = 1.00 + 1.00/X$$

E 7-6.
   Using the following combination of points:

$$1 + 2 + 3;\ 4 + 5 + 6;\ 7 + 8 + 9 + 10;$$

the method of averages based equation is

$$Y = 0.75 + 2.167X + 0.0873X^2$$

E 7-7.
   Least squares fitted equation is

$$Y = 0.70 + 2.12X + 0.119X^2$$
$$\Sigma r^2 = 0.2950$$

CHAPTER 8

E 8-1.
  a: 1/6. b: 5/6

E 8-2.
  1/52

E 8-3.
  With card replacement after each drawing:

$$4/52 \times 4/52 \times 4/52 \times 4/52 = 1/\, 28,561$$

  Without card replacement after each drawing:

$$4/52 \times 3/51 \times 2/50 \times 1/49 = 24/6,497,400 = 1/270,725$$

E 8-4.
  Trimmed mean   = 12.29
  Median   = 12.3
  Arithmetic mean   = 12.275

E 8-5.
Appears to be a random distribution around a positive slope.

E 8-6.
  On regression of y with respect to x, the order of measurement:

$$y = 11.406 + 0.08278\, x$$
$$\text{s.d. of slope} = 0.0228$$
$$\text{s.d. of intercept} = 0.0750$$

slope/s.d. slope = 3.6 which is significant at 95% level of confidence.

E 8-7.

Expected number of runs = 13
s.d. of runs = 1.795
Runs found = 11
$13 - 11/1.795 = 1.11$, not significant at 95% level of confidence

E 8-8.

$MSSD = 0.5152$, $s^2 = 0.5693$
$MSSD/s^2 = 0.905$, significant at 99% level of confidence but not at 99.9% level.

E 8-9.

	A	B	C	D	E
$\overline{X}$	5.34	5.36	5.62	5.49	5.27
s	.202	.240	.252	.126	.307
s^2	.0406	.0575	.0636	.0159	.0943
$s^2/$	.00812	.0115	.01272	.00318	.01888
n					
W	123	87	79	314	53

Critical value for F = 9.6; 2 95% level of confidence
Largest ratio: $.0943/.0159 = 5.9$, not significant
By Cochran test $.0943/.2720 = .347$ compared to .544, hence largest variance is not significantly larger than that of group.

Applying the procedure of Chapter 8:

$$s_c = .0603 \quad q_{1-\alpha} \text{ for 95\% confidence} = 4.23$$

$w = (4.23 \times .0603)/\sqrt{5} = .114$, both C and D differ from smallest E by more than this amount. Comparing the means using the Grubbs test and the Dixon test indicate no outliers. Combining data, and weighting according to precision:

$$\overline{\overline{X}} = \frac{5.37\times123 + 5.36\times87 + 5.62\times79 + 5.49\times314 + 5.27\times53}{123 + 87 + 79 + 314 + 53} = 5.44$$

s^2 of $\overline{\overline{X}} = 0.039$ 95% confidence interval $= 0.016$

E 8-10.

Sign of difference is not important.
a. $2 = (1.96 + 1.64)^2 \times \sigma^2/5^2$
      $\sigma = .196\%$ RSD
b. $n = (1.96 + 1.64)^2 \times .25^2/.5^2 = 3.24$ or 4

E 8-11.

a. .047, .07, .09, .097, .097, .099, .10, .100, .106, .11, .11, .11, .115, .12, .12, .121, .13, .13, .137, .14, .14, .18, .32, .328
b. Plot appears to be essentially symmetrical
c. Median = .1125
d. 25% Winsorized trimmed mean = .115
e. 25% Winsorized s.d. estimate = 0.0225
f. Mean is not significantly biased
g. 95% confidence band is .067 to .157; labs 4, 8, 16, and 18 are outside of this band
h. 99.7% confidence band is .045 to .180; labs 8 and 18 are outside of this band.

INDEX

Accuracy 4, 22, 244
Actual variances 63
Amount of data 13
Analysis of variance (ANOVA) 56, 244
Analytical chemists 6
Arithmetic mean 26, 48, 73, 87
Assignable cause 244
Attribute 244
Attribute data 29, 245
Average deviation 102, 188, 245
Average value 25, 48, 73, 87

Bar charts 123
Basic probability concepts 185
Benefit considerations 15
Best two out of three 209
Bias 2, 3, 4, 5, 12, 22, 105, 245
Biased conclusion 48
Biased sample 48
Bimodal distribution 30
Biometrology 6
Blank 221
Blunders 23, 25
Bounds for the population 83
Boxplot 109, 245

Calibration 4, 6, 38
Causal relationship 208
Censoring 159, 245
    Interval 159
    Left 159
    Non-informative 159
    Right 160
    Type I 160
    Type II 160
Central limit theorem 35
Central line 190
Central value 25, 187
Chance cause 245
Chance fluctuations 190
Chance variability 23
Charts 123
Chemical metrology 6

Chemometrics 11, 245
Chi square 155
Cochran test 55, 107, 242
Coded variances 63
Coding 66
Coding operations 66
Coefficient of determination 208
Coefficient of variation 49, 245
Comparability 14
Comparison of averages 194
Comparison with precision 81
Completeness 14
Computations 66
Computers 7
Confidence band for a fitted line 211
Confidence interval 73, 80, 85, 87, 91,
    149, 150, 154, 157, 245
Confidence interval for estimates of
    standard deviations 80
Confidence interval for a mean 73
Confidence interval for a proportion 149
Confidence interval for difference in propor-
    tions 154
Confidence level 72
Confidence limits 76
Confidence for a fitted line 211
Confidence region for constants of a fitted
    line 215
Conflicts of interest 6
Consensus value of parameter 112
Consensus values 114
Constant dispersion 25
Continuous data 4
Control chart data 221
Control chart limits 222
Control limit 93, 222
Control of measurement process 104
Control sample 223
Correlation coefficient 206, 245
Correlations 39
Cost considerations 15
Cost-benefit considerations 13

Counting data 3
Counting of radioactive disintegrations 3
Critical value approach 200, 246
Critical region 202
Crude t test 101
Cyclic data 191

Data 1
Data analysis techniques 72
Data compatibility 6
Data dredging 231
Data pedigree 11
Data points 143, 246
Data quality 5, 13
Data Quality Indicators 13
Data quality objectives (DQOs) 15
Datasets
    Acid 90
    Cities 130
    Crankshaft 176
    Defects 128
    Furnace 140
    Furnace Temperature 78
    Material Bag 63
    Radiation 42
    Reliable 162
Decisions 1, 21
Decoding 66
Degrees of freedom 52, 246
Dependent variable 128, 246
Descriptive statistics 20
Design of experiments 95, 246
Design of measurement programs 195
Deviation 246
Discrete data 4
Dispersion of data 23
Distribution 247
Distribution-free approach 85
Distribution-free tolerance intervals 85
Distributions 116
    Bernouli 116
    Binomial 116
    Double Exponential 188
    Exponential 170
    Gamma 116
    Gaussian 27, 29, 247
    Hypergeometric 116
    Integer 116
    Log normal 28, 29, 170, 171
    Normal 27, 116, 170, 210

    Poisson 116
    Uniform 29
    Weibull 170
Distributions of standard deviation of the
    mean 51
Dixon test 102, 239
Drifts 16
Duplicate measurements 52, 112
Duplicates 54

Effective number of degrees of freedom 89
Empirical power series 134
Empirical relationships 132
Engineering metrology 6
Engineering tolerance level 114
Enumeration 3
Environmental data 6
Equations solved 135
Erroneous value 24
Error 22, 247
Error bars 127
Estimates of standard deviation 91, 233
Ethical problems of peer review 6
Evaluation of measurement data 231
Excellence in measurement 6
Expected number of runs 190
Experimental data 3
Experimental design 3, 10
Experimental plan 11
Experimental program 12
Experimentation 2
Extreme value data 220
Extreme values 188
Extreme values for variance 107, 241
Extreme values of variance 93
Eye estimate 37, 102, 133, 247

F test 37, 81, 89, 107, 127, 223
False negative decisions 195
False positive decisions 195
Finite population 247
Fitted line 190
Fraction of acceptable items 148
Frequency distribution 27, 124, 210
Functional relationship 208

Gauss 27
Gaussian distribution 247
Gaussian law error 143

Generalized parabola 133
Geometric mean 26, 187
Goodness of fit 33, 127, 208
Grand average 55
Graphed line, deviation from 100
Graphs 126
   Linear 126
   Non-linear 127
   Nomographs 128
Grubbs test 104, 240

Hard data 2
Harmonic mean 194, 247
Hidden replication 54
Histograms 31, 124, 210, 247
Huge error 101
Hypothesis 12, 196

Imprecision 24
Inaccuracy 24
Inaccurate value 24
Independence 36, 38
Independent observations 143
Independent points 36, 52
Independent variable 131, 247
Individuals 247
Inductive statistics 20
Infinite population 247
Information 1
Initial plotting 143
Instability 4, 16
Interlaboratory collaborative testing 112

Joint confidence ellipse 215

Kaplan Meier 161, 248
Kinds of data 22
Kurtosis 39

Laboratory precision 112
Lack of control 4
Least squares 137
Level of detection 30
Limiting mean 24, 248
Limits of detection 82
Limits of uncertainty 14, 127
Linear empirical relationships 132
Linear graphs 126
Log normal distribution 26, 30

Log normal statistics 64
Logarithmic transformation 64
Log rank 165, 248
Long-term standard deviation 221
LOTUS 1-2-3 124

Macro 224, 248
Malfunctions 101
Material variance 56
Mathematical equation 131
Mean 48, 73, 87, 187
Mean square of successive differences
   (MSSD) 192
Meaning of a confidence interval 76
Measure of the association between vari-
   ables 206
Measured value 22
Measurement process 4, 5, 11, 22, 101
Measurement system 223
Measurement variability 126
Measurement variance 56
Median 26, 187, 248
Method Limit of Detection (MDL) 82
Method of averages 134, 248
Method of least squares 137, 248
Method of selected points 133
Metrologists 2
Metrology 6
Middle value 187
Midrange 187
Minimum sample 152
MINITAB 8, 39, 248
Misuse of statistics 152
Mode 26
Moving Average 180
Multimodal distribution 30

Natural data 2
Natural variability 4, 72
Nested design 62
Nomographs 128
Nonlinear empirical relationships 133
Nonlinear graphs 127
Nonparametric tests 217
Nonrandomness 189
Normal distribution 27, 124, 187, 210,
   248
Normally distributed population 192
Normal probability plot 248

Normality 30
Null hypothesis 37, 72, 196
Numerical data 2
NWA QUALITY ANALYST 132
NWA STATPAK 132

Observation 248
Observers, training of 3
Offset 249
One-way analysis of variance 194
Operating characteristic (OC) curve 207, 248
Operator variance 68
Outlier elimination 188
Outliers 100, 192, 249
Outlying laboratories 105
Outlying results 102, 104

P value approach 200, 249
Peer performance 14, 105
Peer review 6, 16
Physical metrology 6
Pie chart 123
Planning 10, 195
Point of inflection 27
Poisson distribution 249
Polishing 188
Pooled standard deviation 52, 233
Pooled statistical estimates 55
Population 47, 249
Population bounds 82
Population mean 102
Population proportion 148
Population statistics 68
Populations 5, 47, 249
Power of the statistical test 195, 249
Power series 133
Precision 22, 76, 105, 249
Precision control charts 223
Predictions 1
Preliminary graphical fit 192
Presentation of data 122
Probability 185, 249
Probability plot 32
Probability plotting paper 32
Probable error 249
Product limit 161, 249
Proficiencies 112
Proficiency testing 113

Propagation of error 94, 249
Property control charts 221
Proportional hazards 165, 250
  Cox 165

Quadratic equation 134
Quaesitum 250
Qualification of laboratories 15
Qualitative data 2
Qualitative identification 13
Quality assessment 15
Quality assurance 15, 221
Quality control 15, 76
Quality of data 13
Quantitative accuracy 14

R chart 223
Random data 36
Random error 23, 250
Random numbers 114, 243
Random sample 48, 250
Randomization 12, 117
Randomly distributed points 36
Randomness 36, 189
Randomness of residual plots 134
Range 26, 54, 223, 250
Range chart 223
Range to estimate standard deviation 54
Ranking 31, 100
Ratio of standard deviation to difference 199
Real world significance 78, 231
Reference materials 112
Relative range chart 224
Relative standard deviation 50, 250
Relevant samples 11
Reliability 5
Repeatability 113, 250
Replicate measurements 16
Reporting statistics 65
Representative samples 5
Representativeness 14
Reproducibility 113, 250
Residual plots 133, 136, 140
Residuals 102, 133
Reviewers 6
Risk of false rejection 104
Round robin tests 105
Rounding 4, 66
Ruggedness test 250

Rule of the huge error 101
Runs 190, 224

Sample
    Definition 47, 251
    Measurement data 5
Sample size 152, 157, 251
Sampling 6
Sampling problems 84
Science of statistics 3, 19
Selectivity 5
Shewhart control charts 221
Short-term drifts 191
Shortcut procedures 216
Significance 78
Significance of apparent trends 191
Significant difference of means 77
Significant figures 4, 6
Simulation 224, 251
Skewness 38
Slope 191
Smoothing 177
Soft data 2
Spike 221
Stable data 36
Stable variance 16
Standard deviation
    Confidence interval 80
    Definition 49, 251
    Distribution 187
    Estimation 48, 52, 54, 188
    Fitted points 102
    Of the mean 49, 251
    Population 28, 48, 73
    Random error 22
    Relative 49
    Of the slope 191
Standard error of the mean 251
Statistical control 15, 25, 75, 209, 231
Statistical planning 10
Statistical power 10, 195
Statistical probability 187
Statistical significance 78
Statistical tolerance interval 82
Student t variate 74, 81, 194, 235
Survival data 159, 251
Symmetrical distribution 27, 217
Systematic departures 132
Systematic error 23, 251

Systematic measurement 15
Systematic trends 224

Tables 122
Theoretical relationships 131
Time series data 174
Tolerance interval (engineering) 252
Tolerance interval (statistical) 82, 252
Tolerance intervals 83
Traceability 11
Training 3
Transformed data 127, 132
Trends/slopes 191
Trimmed mean 188
True value 22
Trueness 252
Two proportions, significance of appar-
    ent differences 153
Type I error 195, 252
Type II error 195, 252

Unbiased data 1
Unbiased information 48
Unbiased measurement systems 25
Uncertainty 14, 19, 88, 252
    Of data 22
    Of estimates of standard deviations
    213
Unevaluated data 2
Uniform distribution 29
Universe 252
Unweighted mean 188

Valid statistical analysis 189
Validation of methods 112
Validity, basic requirements for 36
Variability 4, 13, 23, 26, 33, 55, 80, 128
Variable data 252
Variance 252

Weighting 253
Wilcoxon signed-rank tests 217
Wild data 192
Winsorizing 188

Youden plot 253
Youden test 105, 241

Z-factors 2